# 27th Annual Saudi-Japan Symposium on Technology in Petroleum Refining and Petrochemicals 2017



Dhahran, Saudi Arabia
20 – 21 November 2017

ISBN: 978-1-5108-5513-7

**Printed from e-media with permission by:**

Curran Associates, Inc.
57 Morehouse Lane
Red Hook, NY 12571

**Some format issues inherent in the e-media version may also appear in this print version.**



Printed by Curran Associates, Inc. (2018)



King Fahd University of Petroleum & Minerals
Dhahran 31261
Kingdom of Saudi Arabia

Phone: +966 (13) 860 0000

info@kfupm.edu.sa

**Additional copies of this publication are available from:**

Curran Associates, Inc.
57 Morehouse Lane
Red Hook, NY 12571 USA
Phone: 845-758-0400
Fax: 845-758-2633
Email: curran@proceedings.com
Web: www.proceedings.com

ارامكو السعودية
Saudi Aramco

النَدوة السنوية السعودية اليابانية السابعة والعشرون

*27th Annual Saudi-Japan Symposium*

# التقنية في تكرير البترول والبتروكيماويات

# Technology in Petroleum Refining & Petrochemicals

# PROCEEDINGS: 27

٢-٣ ربيع الأول ١٤٣٩هـ

**November 20-21, 2017**

**KFUPM Research Institute**

**King Fahd University of Petroleum & Minerals**

**Dhahran, Saudi Arabia**

# Preface

This is the Proceedings of the 27th annual symposium "*Technology in Petroleum Refining & Petrochemicals*" held on November 20-21, 2017 at the Research Institute of King Fahd University of Petroleum & Minerals (KFUPM), Dhahran, Saudi Arabia. This event, jointly organized by KFUPM, Saudi Aramco, the Japan Petroleum Institute (JPI), and Japan Cooperation Center, Petroleum (JCCP), brings together researchers from industry and academia to discuss technologies related to refining and petrochemicals.

At this year's symposium, there are 20 presentations in six sessions featuring papers on, catalytic reactions, hydroprocessing, aromatics,, catalytic olefins, cracking, dehydrogenation. Among the distinguished speakers from Japan, we have experts from the University of Tohoku, Yokohama National University, JXTG, JGC and Kyoto University. For Local participation, there are distinguished speakers from Saudi Aramco R&DC, Honeywell UOP, SABIC R&D, KFUPM/CRP as well as chair professor from Zaragoza University.

We extend our thanks to KFUPM, Saudi Aramco, JPI, JCCP, distinguished speakers, and attendees for their active participation and look forward to your participation in our next symposium.

***Dr. Sulaiman S. Al-Khattaf***
Director, Center for Refining & Petrochemicals
***http://ri.kfupm.edu.sa/refining-petrochemicals/***
KFUPM-Research Institute, Dhahran 31261, Saudi Arabia
Email: skhattaf@kfupm.edu.sa

# Contents


| No. | Title/Speaker | Page |
|---|---|---|
| 1. | Vision of the Japanese oil industry and importance of the education of human resources, *Jun Mutoh, JPI, JXTG (KEYNOTE)* | 4 |
| 2. | Commercialization of High-Severity Fluid Catalytic Cracking (HS-FCC) process for light olefins production in refineries, *Iwao Ogasawara, JXTG Energy, Japan* | 5 |
| 3. | A case study of HS-FCC reactor using kinetic modeling, *Qi Xu, Saudi Aramco R&DC* | 13 |
| 4. | Degradation of heavy-duty coating applied to inner bottom plates of oil storage tanks and its evaluation method based on electrical impedance analysis, *Shinji Okazaki, Yokohama National University, Japan* | 15 |
| 5. | Applications of process and kinetic models to optimize the operation of hydrocracker, *Rashid Ansari, Saudi Aramco R&DC* | 27 |
| 6. | Effect of acid medium and initial Si/Al ratio on the synthesis of mesoporous materials with enhanced acidity and hydrothermal stability, *Lianhui Ding, Saudi Aramco R&DC* | 35 |
| 7. | Kinetics of simultaneous HDS of DBT and 4-MDBT/4,6-DMDBT over CoMoP/$\gamma$-$Al_2O_3$ catalysts, *Syed A. Ali, KFUPM/RI* | 37 |
| 8. | Development of a supercritical water cracking process for pipeline transportation of extra heavy crude oil, *Takayoshi Fujimoto, JGC Corporation, Japan* | 38 |
| 9. | Hydrodynamics and solids mixing in fluidized beds with inclined-hole distributors, *Alhussain Bakhurji, Saudi Aramco R&DC* | 50 |
| 10. | Zeolite-based catalyst performance for catalytic reforming, *Mohammad Al-Rebh, Saudi Aramco R&DC* | 53 |
| 11. | Petrochemicals – technical and economic challenges for this century, *Patrick Whitchurch, Honeywell UOP, USA (KEYNOTE)* | 56 |
| 12. | Catalytic copolymerization of $CO_2$ and diols using $CeO_2$ and 2-cyanopyridine, *Keiichi Tomishige, Tohoku University, Japan* | 57 |
| 13. | Toluene Methylationto produce para-Xylene using ZSM-5 catalysts , *Arudra Palani, KFUPM-RI; Ali Jahel, Honeywell UOP* | 69 |
| 14. | Metathesis of 2-pentene over Mo and W supported mesoporous molecular sieves, *M. Naseem Akhtar, KFUPM-RI* | 71 |
| 15. | $CO_2$ fixations with homogeneous catalysis, *Yasushi Tsuji, Kyoto University, Japan* | 72 |
| 16. | Catalyst development for hydrogenation of $CO_2$ for energy storage, Ali Al-Jishi, *Saudi Aramco R&DC* | 84 |
| 17. | Size-controlled synthesis of gold nanoparticles on fibrous silica nanospheres (KCC-1) and their activity in the oxidation of CO, *Ziyauddin Qureshi, KFUPM-RI* | 86 |
| 18. | Bifunctional selective hydrosilylation of terminal alkynes by bis(carbene) iridium(III) and rhodium(III) complexes, Luis *Oro, Zaragoza University, Spain and KFUPM* | 87 |
| 19. | Process Intensification, a promising approach in separating a ternary system using distillation to reduce energy consumption, *Dafer Al-Shahrani, Jagan Rallapalli, SABIC R&D, Riyadh* | 88 |
| 20. | Thermo-neutral reforming of diesel for auxiliary power unit (APU) application, *Mohammed Albuali, Saudi Aramco R&DC* | 89 |

Paper No. 1 (keynote)

# Vision of the Japanese Oil Industry & Importance of the Education of Human Resources

***Jun Mutoh***
*Representative Director, Executive Vice President, JXTG Holdings, Inc.*
*President, Japan Petroleum Institute*

The JXTG Nippon Oil Group, an energy company group that supplies about 50% of Japanese oil demand, was established through a merger between the JX Nippon Oil Group and the TonenGeneral Group in April 2017.

Global oil demand is growing steadily, led by Asia Pacific, but in Japan, demand is decreasing year by year. The Japanese petroleum industry has implemented company integration and reorganization to improve productivity, efficiency and competitiveness amid industry deregulation and declining demand.

Established in this environment, JXTG is building a global supply system, and is shifting to higher value added petrochemicals. We are also focusing on electric power, LNG, hydrogen and renewable energy as future businesses.

The keys to the success of these businesses are advanced technology and human resources.

Saudi Arabia and Japan have built a strong relationship. In addition to business transactions, we have cooperated in the technological advancement of the petroleum industry and the education of human resources. In the future, it will be increasingly important to expand technological cooperation in the industry and foster human resources skilled in next-generation technology. I anticipate further mutual development through the activities of JPI and JCCP.

Papar No. 2

# Commercialization of High Severity Fluid Catalytic Cracking process for light olefins production in refineries

***I. Ogasawara***
*JXTG Nippon Oil & Energy Corporation, Japan*

The Fluid Catalytic Cracking (FCC) process has undergone a long evolution of hardware and catalyst changes from bed cracking with amorphous catalyst to short contact time riser cracking with sophisticated zeolite catalyst systems. Improvements to the process have provided a wide degree of flexibility to selectively target production of distillates or gasoline or propylene from VGO and residue feeds thereby making the FCC the most widely used conversion process.

More generally, the objective of the process is to produce high valued products and increasingly this includes fuels and petrochemicals, including light olefins and aromatics. At present, over 30% of the worldwide propylene supply comes from FCC related processes (FCC, RFCC etc). Fluctuating product demand and price have caused most new project developers to demand product flexibility for long term profitability and process integration with petrochemical facilities for added synergy and cost savings.
In order to respond to these market demands, a new High Severity FCC (HS-FCC) process has been developed by an alliance of King Fahd University of Petroleum & Minerals (KFUPM), Saudi Aramco and JXTG Nippon Oil & Energy (JXTG), culminating in a 3,000 BPD semi-commercial unit in operation in Japan for 2011-2013. The process provides high light olefin yield from a wide variety of feedstocks utilizing high-severe reaction conditions, a novel down flow reaction system and proprietary catalyst. HS-FCC is now available for license from a Global Alliance by Axens and TechnipFMC Process Technology.

Figure 1. HS-FCC Semi-Commercial Unit

## 1. Features of HS-FCC

FCC utilizes acidic zeolite catalysts to crack heavy hydrocarbons into lighter fuels such as gasoline and distillate, and under more severe conditions into lighter olefins such as propylene and butylene (and to a lesser extent ethylene). Complex secondary reactions that can degrade the primary products to less valuable components should be limited to retain product selectivity and refinery profitability. For HS-FCC, the objective is to not only improve the selectivity for normal fuels production, but also to maximize the potential of light olefin and petrochemical production at high severity conditions. HS-FCC provides a total system to maximize product selectivity and, in particular, propylene yield. Three key elements are required to attain this objective:

- Highly selective catalyst and additive system;
- Optimized reaction conditions; and
- Down flow, short contact time reaction system with rapid catalyst separation.

The balance of these elements and realization at commercial scale is the key to success.

**2. Catalyst system**

The catalytic cracking reaction pathways are complex, with primary formation of olefinic products and parallel bi-molecular hydrogen transfer reactions leading to paraffin formation and aromization of naphthenes. Managing the acid site density of the catalyst can suppress hydrogen transfer and isomerization reactions to maximize olefins production.

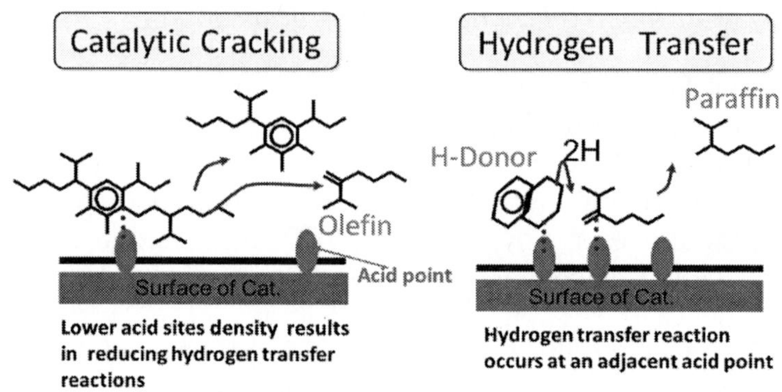


Figure 2: Concept of HS-FCC catalyst

The HS-FCC catalyst uses a high USY zeolite content system with very low acid site density catalyst formulated to minimize hydrogen transfer reactions for high olefin selectivity and low coke and gas selectivity. Figure 2 illustrates the concept of HS-FCC catalyst.

**3. Optimized Reaction Conditions**

When targeting maximum petrochemicals production, HS-FCC operates under more severe conditions than conventional FCC. The main reaction conditions applied and the advantages and challenges presented are summarized below:

| | Advantages | Challenges |
|---|---|---|
| High Temperature | High conversion & olefins selectivity | Increased thermal cracking, product degradation |
| Short Contact Time | Reduced secondary reactions & thermal cracking | Reduced conversion, rapid mixing and separation required |
| High Catalyst / Oil | Increased catalytic cracking | High catalyst circulation, uniform flow, mixing, & separation |

High reaction temperature coupled with short contact time increases the primary reactions towards olefins while limiting the undesired secondary reactions of hydrogen transfer and thermal degradation. A consequence of the increased severity and short contact time the need for higher catalyst circulation (Catalyst to Oil mass ratio or C/O) to provide the required heat to the reactor and sufficient catalyst activity to achieve high conversion at short contact time. The range of operating conditions for a conventional FCC and HS-FCC are summarized in Table 1.

Table 1. Typical operating conditions for FCC and HS-FCC

| | FCC | HS-FCC |
|---|---|---|
| Reaction T, C | 500-550 | 550 – 650 |
| Contact Time, s | 2 – 5 | 0.5 – 1.0 |
| Catalyst / Oil, wt/wt | 5 - 8 | 20 – 40 |
| Reactor Flow | Up Flow | Down Flow |

## 4. Down Flow Reaction (DFR) System - Hardware

The specific reaction conditions with very high C/O result in certain challenges in a conventional up flow FCC riser reactor system where the catalyst required for the reaction is lifted up the reactor pipe or riser by the vaporized and cracked hydrocarbon feed. In up flow fluid-solid systems, the solids or catalyst are conveyed upward against the force of gravity by drag forces from the rising gases (hydrocarbons). As a result, all riser reactor systems have varying degrees of catalyst back-mixing and reflux along the walls, particularly in the feed injection or catalyst pickup zone at the bottom of the riser reactor. Consequently, reactor residence time varies from center of the riser to walls, and it is hard to realize and manage the short contact time within less than 1 second. At very high C/O significant back-mixing is unavoidable. This problem is overcome in a down flow reactor (DFR) where both the catalyst and feed flow downward together.

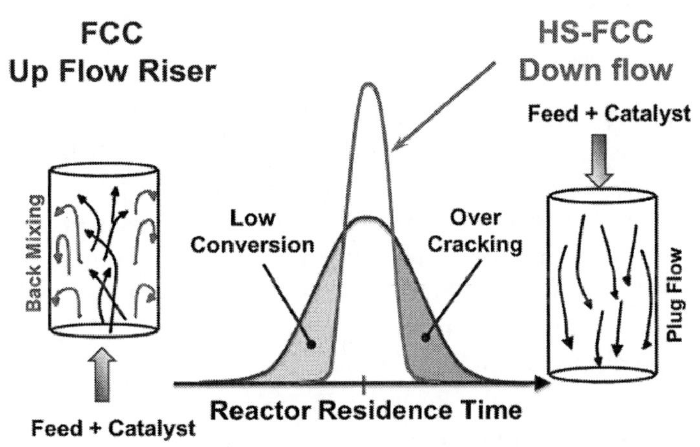


Figure 3: Up Flow vs Down Flow residence time profiles

Down flow fluid-solid reaction systems have been of increasing interest in recent years to achieve plug flow reaction conditions as summarized Cheng[2]. When plug flow conditions are achieved, more selective primary cracking results in greater selectivity. FCC pilot work demonstrating the effects of short contact time and down flow have been reported by Del Poso[3] and Abul-Hamayel[4] as shown in Figures 4 a-b. The general trend reported is that of greater gasoline selectivity at short contact time down flow with a maximum yield achieved at a higher conversion level. This effect is seen in Figure 4a where the maximum gasoline yield is about 5 wt% higher in the down flow system. When olefins are of interest, the more selective down flow reaction environment can produce substantially more light olefins (C2-C4) at the same gasoline yield compared to a conventional up flow system (Fig. 4b).

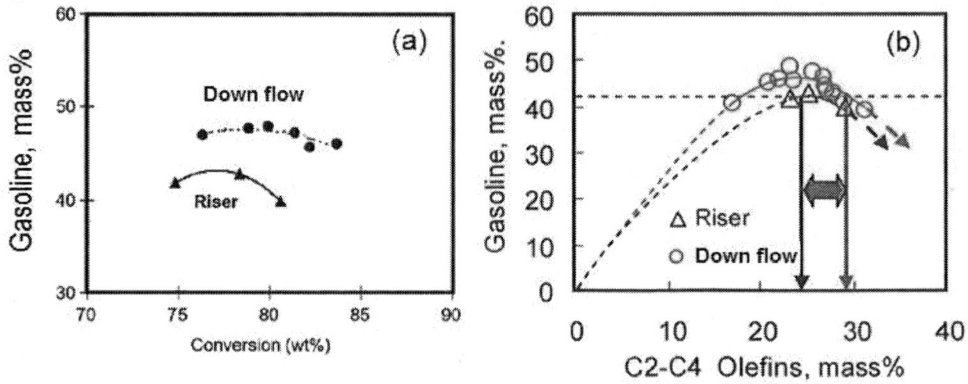


Figure 4: Selectivity benefits of a down flow reaction system (4)

Although the idea of controlled high severity, short contact time down flow reaction, has been considered for some time, achieving this successfully at commercial scale has been elusive. Extensive pilot work at the 0.1 BPD scale demonstrated the principle, catalyst system and operating conditions, but did not address how rapid mixing, reaction and efficient catalyst/gas separation can be achieved at large scale with a target residence time on the order of 0.5 sec. At commercial scale, equipment design for very short contact time with the mechanical integrity to withstand high velocity catalyst circulation in a coking environment requires extensive research, development, and demonstration.

## 5. R&D History

The challenges of developing this new technology required a systematic research & development program undertaken by KFUPM, Saudi Aramco and JXTG, with supports of Japan Cooperation Center, Petroleum (JCCP).Early pilot work by both KFUPM and JXTG in 1996-2000 demonstrated the benefits of high severity operation at controlled short contact time in a down flow reactor. Aramco became an active participant in the scale up effort to design a 30 BPD demonstration unit. JX conducted large scale, 30 BPD equivalent, cold flow testing of the catalyst circulation loop and reactor-separator equipment to validate the design of the demonstration unit.

The demonstration unit shown in Figure 5 was operated from 2003-2004 at the Aramco Ras Tanura refinery. Results from the demonstration unit validated the HS-FCC concept with good agreement of scaled-up between 0.1 BPD pilot results and the 30 BPD demonstration as shown in Figure 6.[5, 6]

Desulfurized VGO was cracked at high severity in both the pilot and demo units. A very high propylene yield over 10% was obtained along with a very high octane gasoline.

Figure 5: HS-FCC Demonstration unit

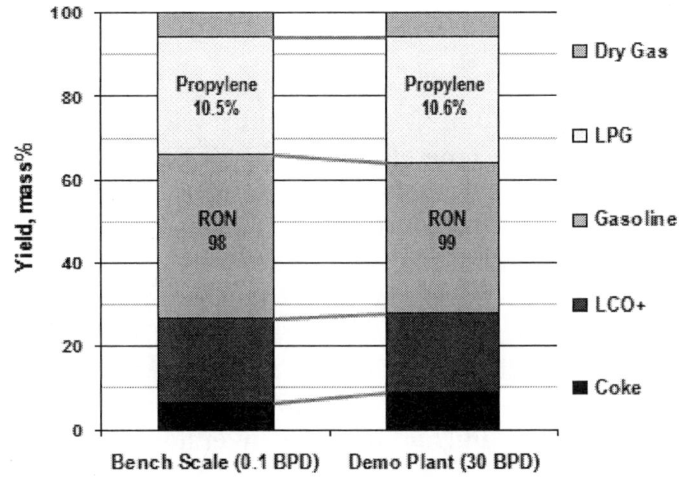


Figure 6: Bench scale vs Demonstration scale results on desulfurized VGO at high severity conditions

Work immediately began on scale-up to commercial unit. Important lessons were learned concerning equipment design and larger scale cold flow work was undertaken by JXTG in Japan at the 500 BPD equivalent scale to optimize feed injection zone and separator design – Figure 7. This work was coupled with CFD simulations to assist in larger scale equipment design[6].

Figure 7: 500 BPD equivalent cold flow testing to scale-up and optimize reaction system

## 6. Semi-Commercial Unit

With the successful demonstration of the HS-FCC technology at the 30 BPD scale completed, it was time to look forward to scale-up to a full size commercial unit and plan for future licensing of the technology. Several FCC licensors were interviewed and evaluated before Axens and TechnipFMC Process Technology were selected to assist in the design of a 3,000 BPD semi-commercial unit, plan for a larger commercial unit, and serve as exclusive licensor for the HS-FCC technology, relying on their extensive knowledge in FCC and RFCC design.

A complete 3,000 BPD HS-FCC unit with main fractionator, gas plant, and flue gas treatment was designed for the JXTG Mizushima refinery. Design works of the unit started in 2007, and was put on-stream in early 2011. The unit was operated in 2011 – 2013 (Figure 1). Performance trials were conducted to evaluate yields and product properties for widely different feeds and to demonstrate equipment reliability. The results showing yields for several feeds of VGO, Hydrocracker (HDC) bottoms and Atmospheric Residue are shown in Table 2. Combined light olefins (C2-C4) yields of 34 to 40 wt% have been demonstrated with 17 to 20 wt% propylene and 4-5 wt% ethylene. The yield of butenes is enough high to offer opportunities for greater petrochemical integration, including oligomerization and/or olefin conversion configuration for even higher propylene production.

| | **HDT VGO** | **VGO + HDC BTM** | **HDT AR** |
|---|---|---|---|
| Feed SPGR | 0.86 | 0.85 | 0.92 |
| Reactor T, ºC | 600 | 575 | 600 |
| Light Olefins, w% | 40 | 39 | 34 |
| C2= | 4 | 4 | 5 |
| **C3=** | 20 | 19 | **17** |
| C4= | 16 | 16 | 12 |
| C5-220 Gasoline, w% | 34 | 35 | 31 |
| RON | 96 | 98 | 98 |

Table 2: Semi-Commercial Unit Performance

When viewed from a petrochemicals perspective, the ethylene produced becomes a significant boost to the economics. The gasoline also has value beyond fuels with an octane of 96-98, olefin content of 25 to 40 wt%, and 35 to 50 wt% aromatics.

With a controlled short contact time, high C/O and plug flow reaction system, HS-FCC is well adapted to be highly selective for both light and residue feed conversion to petrochemicals. Throughout the program equipment evaluation, inspection and reliability data were gathered to guide further development and scale-up to a fully commercial scale of at least 30,000 BPD. In parallel to this work, CFD simulation of the Down Flow Reactor (DFR) and separator hydrodynamics are being combined with a kinetic model to analyze the results, validate the kinetic models, and enable accurate predictions at commercial scale for future feeds and reactor configurations.

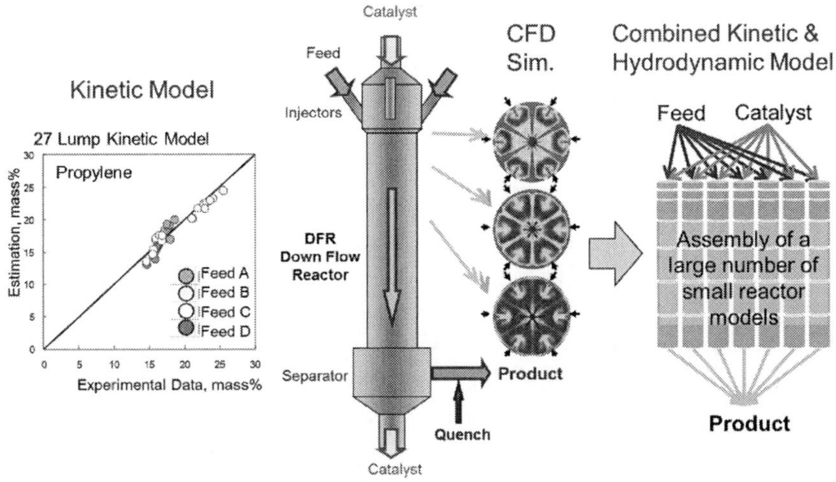


Figure 8: Combined kinetic and hydrodynamic modeling assists design and scale up

## 7. Bottom to Chemical

The HS-FCC process produces a large amount of light olefins and gasoline which contains aromatics as well, from bottom streams in refining, and serves as a key process to integrate refining with petrochemical shown in Figure 9[7]. Total yields of propylene, butene and aromatics from HS-FCC exceed 50 vol% and is higher than the ones of RFCC and Naphtha Cracker shown in Figure10 [7].

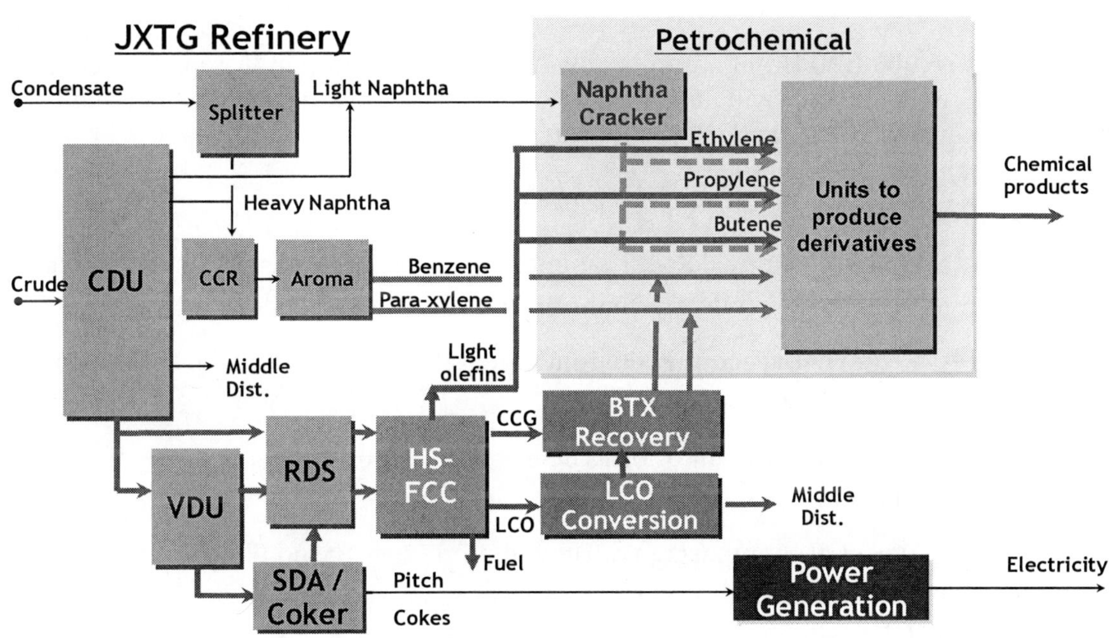


Figure 9: HS-FCC to integrate Refining and Petrochemical

## 8. Go for Propylene

HS-FCC technologies have been completed as results of systematic process research, catalyst development, pilot work, 30 BPD demonstration unit testing, and semi-commercial operation and testing at the 3,000 BPD scale. These successful results and the modeling tools developed for further scale up make the technology ready to install commercial units to maximize light olefins and BTX products.

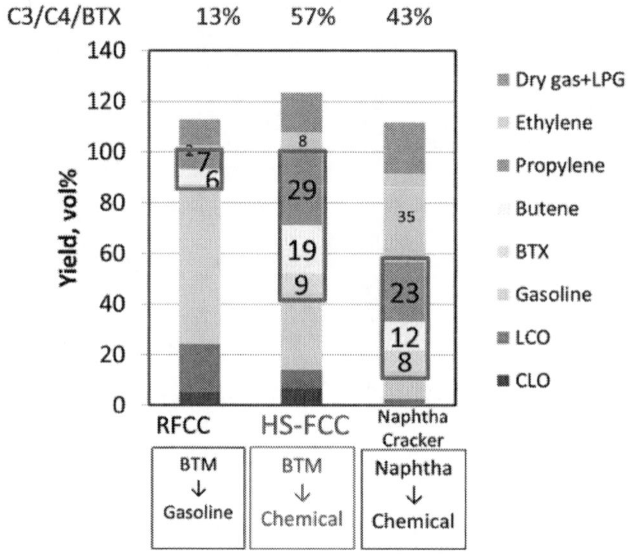


Figure 10: C3/C4/BTX yields comparison

A large scaled 76,000 BPD HS-FCC unit is now under construction in Korea, and scheduled to start-up in 2018. The unit will become a new runner to open the window between refining and petrochemical through residue streams.

## References:

(1) Maghrabi et al., 10th Annual Saudi-Japanese Symposium on Catalysis in Petroleum Refining and Petrochemicals, Dhahran, 2000.

(2) Cheng et al., Downer reactor: From fundamental study to industrial application, Powder Technology, 183, 2008.

(3) Del Poso et al., Development of Ultra Selective Cracking Technology, 2nd IFP and S&W FCC Forum, 1996.

(4) Abul-Hamayel et al., Comparison of Downer and Riser Based Fluid Catalytic Cracking Process at High Severity Conditions: a Pilot Plant Study, Petroleum Sci. Tech., 2004.

(5) Redhwi et al, Meeting Olefins Demand in a Novel FCC Technology, World Petroleum Congress, 2005

(6) Okazaki et al., High-severity Fluidized Catalytic Cracking (HS-FCC) – Go for Propylene!, World Petroleum Congress, Doha, 2011.

(7) Ogasawara, High Severity Fluidized Catalytic Cracking (HS-FCC), World Petroleum Congress, Moscow, 2014.

# Paper No. 3
# A Case Study of FCC Reactor Using Kinetic Modeling

***Qi Xu, Ali Jawad, Musaed Ghrami, Abdennour Bourane***
*Saudi Aramco, Dhahran, Saudi Arabia 31311*

Fluid Catalytic Cracking (FCC) technology, as one of the most important process technologies used in petroleum refineries since the last century, converts heavier feedstocks, usually vacuum gas oil or heavy residue, into more valuable products, such as gasoline, LPG and olefins. A typical FCC unit consists of a riser or downer reactor section, a catalyst stripper section and a catalyst regeneration section. In the FCC reactor, the feedstock comes into contact with the surface of catalyst where the long-chain molecules of the high-boiling hydrocarbons are converted to shorter molecules through endothermic cracking reactions. The amount of energy required for this conversion is normally provided by the heat of combustion, which results from burning coke in the regenerator section. Heat is also needed to vaporize the feed in the inlet to the FCC unit. Hence, the heat balance on an FCC unit determines the basic interaction between the key variables of this unit. Understanding this heat balance helps to better understand the FCC process and to take decisions on how to improve the operation of real FCC unit.

The main objective of this work is to study the heat balance for a typical FCC unit. First, a simulation model was set up in Aspen HYSYS employing a twenty-one lumps reaction network system. This model is calibrated using experimental data and used for the case study. Then, the impact of the reactor temperature on key variables including conversion, catalyst to oil ratio, coke yield, delta coke yield, regenerator air volume flow, and propylene yield is investigated. In the end, the FCC heat balance principles are used to help explain the trends observed in this study. This study also examines how the residence time affects the produce yields.

Modeling results show that the value of reactor temperature has big impact on the catalytic process in the FCC reactor. As the temperature in the reactor goes up, so does the requirement for heat input to meet the demand for heat of vaporization and heat of reaction. Because the solid catalyst serves as carrier for the heat to move from the regenerator to the reactor, it follows naturally that the rate of circulation for catalyst increases, leading to an uptrend in catalyst to oil ratio, as shown in Figure 1A. Subsequently, the cracking reaction rate is enhanced by both the elevated temperature in reactor and the boosted availability of catalyst, causing the total conversion propylene yield to go up, as shown in Figure 1B. As a result of the promoted cracking reactions, the coke yield is increased as well. The delta coke on the catalyst, on the other hand, is decreased. The air flow rate in the regenerator trends higher to provide the necessary oxygen for the combustion of coke, to provide the required heat.

The effects of residence time of the feed in the reactor has also been scrutinized. When the residence time is extended, it is observed that the conversion increases with decreased production of heavy components, as shown in Figure 2. This is accompanied by increased coke yield and light gas production. Interestingly enough, within the setup in this particular simulation model, the gasoline yield experiences a maximum value in the middle, with the yield going up first and then down as residence time increases.

In conclusion, the reactor section of a typical FCC unit was successfully simulated using an FCC reactor model in Aspen HYSYS. The model was validated and employed successfully to investigate the behavior of the FCC system in response to variations in reactor temperature and residence time. The impact of the reactor temperature on key variables were explained using the FCC operation and heat balance principles. The yield slates were analyzed at different residence time in the reactor. This study will help to improve the operation of the existing FCC unit and enhance future reactor design.

Figure 1. Variation of key variables with reactor temperature.

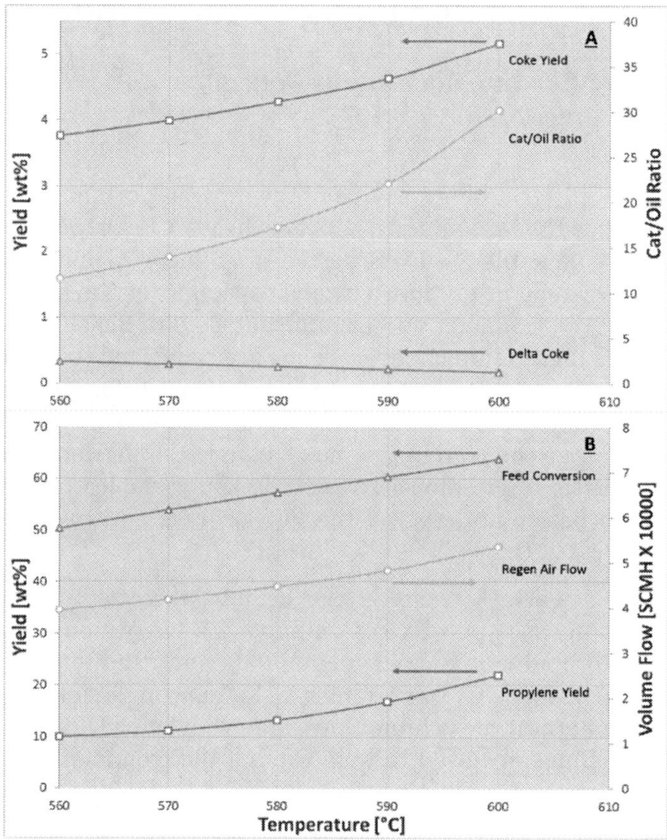


Figure 2. Yield distribution with different residence time.

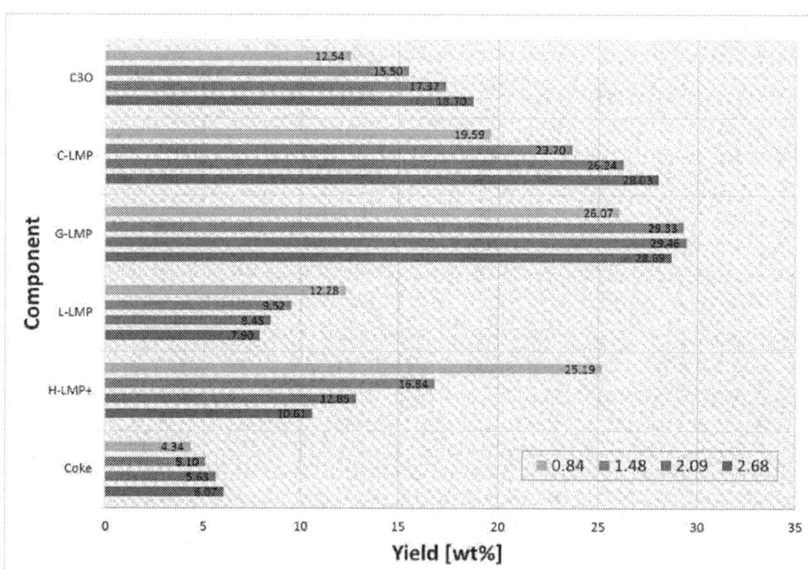

Paper No. 4

# Degradation of heavy-duty coating applied to inner bottom plates of oil storage tanks and its evaluation method based on electrical impedance analysis

***Shinji Okazaki, Koya Tokutake*, Haruki Nishi*, Daisuke Ito, Naoya Kasai Kazuyoshi Sekine,*** *Yokohama National University, National Research Institute of Fire and Disaster**

Inner bottom plates of crude oil storage tanks are generally exposed to mixtures of corrosive drain water and sludge. In order to isolate the steel plates from aggressive environments, high-performance anticorrosion coatings such as a vinyl ester resin organic coating containing glass flakes have been recently utilized. However, the barrier property gradually deteriorates in a long in-service period. In this study, the degradation processes were investigated using acceleration test which imposed the temperature gradient along the thickness direction on coating samples. The glass flake exhibited a good blistering resistive performance and a thick coating system of about 1 mm thickness did not show any sign of blistering in our experimental condition. However, black rust was detected at the interface between coating and steel substrate even in the case of thick coating system. For tracing degradation process in more detail, electrical impedance measurement was adopted and the field data were acquired from actual oil storage tanks. It was found that the impedance behaviors of degraded coating was well explained by using electric equivalent model including CPEs (Constant Phase Elements). The data of extracted CPEs were then analyzed by using extreme value statistics as a tool for diagnosis of heavy-duty coatings.

# Degradation of Heavy-duty Coating Applied to Inner Bottom Plates of Oil Storage Tanks and Its Evaluation Method Based on Electrical Impedance Analysis

**Shinji Okazaki, Koya Tokutake*, Haruki Nishi*, Daisuke Ito, Naoya Kasai and Kazuyoshi Sekine**
**Yokohama National University, 79-5, Tokiwadai, Hodogaya-ku, Yokohama, Kanagawa 240-8501, Japan**
**National Research Institute of Fire and Disaster*, 4-35-3, Jindaiji-Higashimachi, Chofu, Tokyo, 182-8508 Japan**

## Abstract

The steel bottom plates inside crude oil storage tanks are generally exposed to mixtures of corrosive drain water and sludge. In order to isolate these plates from aggressive environments, high-performance anticorrosion coatings, such as vinyl ester resin containing glass flakes, have been recently utilized. However, the barrier property of coatings gradually deteriorates over a long in-service period. In this study, the degradation processes were investigated using acceleration tests that imposed a temperature gradient along the thickness of coated steel specimens. Our experiments showed that the glass flake coating exhibited good blistering resistance and a thick coating of about 1 mm thickness did not show any sign of blistering. However, black rust was detected at the interface between the coating and steel substrate, even in the absence of blistering. To trace the degradation process in more detail, electrical impedance measurements were carried out on actual oil storage tanks. It was found that the impedance behavior of degraded coatings were accurately explained using an equivalent electric circuit model with constant phase elements (CPEs). The data of extracted CPEs were then analyzed using extreme value statistics as a diagnostic tool for heavy-duty coating systems.

## 1. Introduction

In order to secure a stable energy supply and prevent the significant social impact caused by oil supply shortages, stockpiling of crude oil and petroleum products is one of the most important national policies in Japan. At present, it is announced that approximately 88 million kL of petroleum and associated products are stored in Japan, mainly in national and private stockpiles. These stockpiles will enable Japan to maintain everyday life for about 214 days in the case of oil import disruption. Most of the crude oil in these stockpiles is stored in huge oil storage tanks made of steel. The inner surface of the bottom plate of an oil storage tank is normally exposed to a corrosive electrolyte solution called drain water, which is characterized in Table 1. From Table 1, it can be seen that drain water is a neutral-pH solution that contains chloride ions and oxygen in concentrations sufficient to cause corrosion. Since the sulfate ion concentration is smaller than that of sea water, in the range of 800–3,400 ppm [1], it is assumed that the occurrence of $H_2S$ resulting from the reduction reaction of $SO_4^{2-}$ by sulfate-reducing bacteria [2] is almost negligible. Furthermore, it has been reported that the $H_2S$ concentration in drain water is very low, even if the $H_2S$ concentration in the gas phase of the stored product is very high (e.g., when the $H_2S$ concentration in the gas phase was 8,000 ppm, the drain water concentration was 10 ppm) [3]. Coatings were rarely applied to protect the internal surface of the bottom plates of tanks before the 1960s. However, tank overhaul inspections revealed severe bottom plate corrosion because of drain water exposure. Since then, internal coatings have been employed in order to extend the service life of tank bottom plates. In the past, the coatings applied to bottom plates were tar epoxy and polyurethane resin. Recently, a vinyl ester resin organic coating containing glass flakes has seen increased use as a high-performance

Table 1 typical composition of the electrolyte solution in contact with an internal bottom plates of oil storage tanks. (Results of the monthly composition analysis over 1 year.)

| pH | Dissolved oxygen (ppm) | Chloride ion concentration (ppm) | Sulfate ion concentration (ppm) |
|---|---|---|---|
| 6.7–7.3 | 0.1–4.5 | 2485–15760 | 2.0–83.0 |

anticorrosion coating for aggressive environments [4,5], so it has been applied in many tanks.

A coating reinforced with glass flakes has several advantages over a general resin coating. First, the reinforced coating with one-ply application is thicker than a resin coating without flakes. Furthermore, this type of coating shows good strength and exhibits excellent isolation performance against environmental constituents, because the flakes increase the length of the diffusion path for corrosion promoters, such as water, dissolved oxygen, and aggressive ions. The use of high-performance coating systems it is thought to be responsible for the absence of fatal oil leaks over the past 20 years in national oil storage tanks. However, the gradual deterioration of the anticorrosion performance of the coating with age cannot be avoided. Therefore, it is important to accurately predict the residual service life of the coating to prevent serious accidents.

Oil storage tanks are subject to periodic internal inspections according to Japanese regulations. Detecting the physical failure of the coating is often done by visual inspection. Any visually detected coating defects, such as flaws, delamination, and blisters, are repaired. In particular, blisters are frequently regarded as a critical defect, because they signify failure of the coating/steel bond and are accompanied by a corrosion reaction in the metal substrate under the blister [6,7]. Although a visual inspection is simple, the evaluation depends on the inspector's skill, and a decrease in the coating's barrier performance can not be detected visually. In addition, it is necessary to evaluate the anti-corrosion performance of non-defective coatings, because the area of visually observable defects accounts for less than several percent of the entire bottom surface.

Impedance measurement is a non-destructive and quantitative evaluation method that is able to determine the characteristics of organic coatings in detail [8–10]. Therefore, this method has attracted much attention for quantitative evaluation of bottom plate coatings in oil storage tanks. However, impedance measurements are not utilized to evaluate the residual service life for coatings because, presently, only a limited frequency range is used as the indicator of the degradation of the coating. As the first step for the accurate prediction of the residual service life of a coating, it is important to clarify the degradation mechanism with a sufficient interpretation of the electrical properties of the coating. In this study, a field survey using impedance measurements was conducted on an actual oil storage tank that had been in service for a long time. The degradation mechanism of the coating of the tank was investigated.

## 2. Brief introduction of impedance theory

To interpret the physical and chemical properties of electrochemical systems, three simple properties are generally used: resistance ($R$), capacitance ($C$), and inductance ($L$). However, the actual impedance characteristics deviate from the ideal behavior represented by these simple properties because the physical and chemical properties in many cases cannot be regarded as part of a homogeneous system [11]. A constant phase element (CPE) is a useful circuit element to express the non-ideal behavior of electrochemical systems. Impedance spectra can be well-described using a CPE instead of an equivalent RLC circuit. The CPE impedance is defined according to Eq. (1),

$$Z = \frac{1}{T(j\omega)^n} \quad -(1)$$

where $Z$ is the CPE impedance ($Z = Z' + jZ''$), $j$ is the imaginary unit ($j^2 = -1$), $\omega$ is the angular frequency (rad/s), $n$ is CPE power ($-1 \le n \le 1$), and $T$ is a CPE constant ($s^n \Omega^{-1}$). Since the CPE constant $T$ includes the ohm$^{-1}$ dimension, its numerical value is related to the electrical conductivity of the measured system. Then, the value of CPE power $n$ can be expressed using a phase shift $\theta$ (degrees), as denoted in Eq. (2).

$$n = \frac{\theta}{-90^\circ} \quad -(2)$$

For $n = 1$, the CPE indicates an ideal capacitor; for $n = 0$, an ideal resistor is obtained; and $n = 0.5$ defines the Warburg impedance [12].

Although the physical origins of the CPE are still unclear, it is generally agreed upon that the

CPE behavior arises from the distribution of time constants. The result comes from the distributions of the physical properties of inhomogeneous electrochemical systems, such as electrode surface roughness, electrode porosity and current, and potential distribution [13]. Some recent research [14,15] suggests that the CPE behavior of coatings appears to be due to the distribution of time constants as a result of inhomogeneous electrolyte uptake in the coating. Therefore, it is important to confirm the CPE behavior of the coating applied to actual oil storage tanks.

## 3. Experimental

3.1 Preparation of test specimens

Small steel plates (150 × 70 mm, 3.2 mm thick, specification SS400) were used as test specimens. Before being coated, each plate was sandblasted to meet ISO Sa 2-1/2 cleanliness level. A vinyl ester resin without glass flakes was used as primer and applied to the pre-treated steel surfaces to form coating thicknesses of 50, 100, 300, and 500 µm. In order to simulate an actual coating system specification, vinyl ester resin with glass flakes was over-coated to thicknesses of 100, 300, 400, and 900 µm on primer coatings 50 µm thick. Fig. 1 shows a cross-sectional view of a coating system composed of vinyl ester resin and glass flakes. Many thin glass flakes with a length of about 100 µm on average were well-dispersed and aligned horizontally.

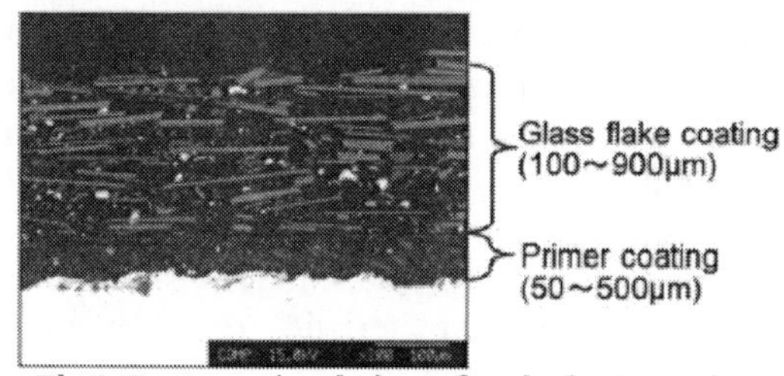


Fig.1 Cross-sectional view of a vinyl ester resin organic coating containing glass flakes on a steel plate.

3.2 Accelerated aging test using temperature gradient method

In order to accelerate the aging of the coating systems, we conducted a temperature gradient test. Fig. 2 illustrates the experimental setup for aging acceleration. The surface of the coating was kept at 40°C by contact with hot water. The back side of the steel plate was cooled to 20°C. Thus, a temperature gradient of about 20°C was imposed along the thickness of the specimen.

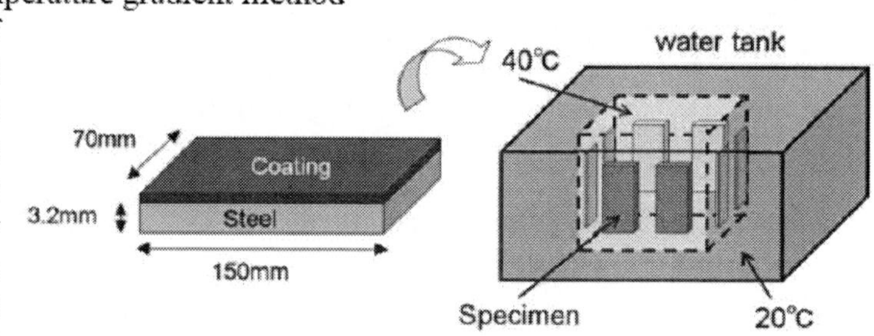


Fig.2 Schematic diagram of test specimens and immersion test under temperature gradient.

3.3 Crude oil storage tank specification and inspection data

The target crude oil storage tank has a capacity of 110,000 kL, and it had stored crude oil for over 266 months. The internal bottom plates of the tank are coated with a vinyl ester resin organic coating containing glass flakes. The coating is composed of three layers: primer, middle coat, and top coat. The base coating solution contains vinyl ester resins (novolac and/or bis-phenol based) mixed with pigments, additives, and thinner (styrene monomer) according to the manufacturer's specification. A typical hardener is a methyl ethyl ketone peroxide. The mixing ratio of the base coating solution and the hardener is 100:0.5–3.0 by weight. The ratio is adjusted according to the application conditions, such as ambient temperature. The middle and top coat include glass flakes in the range of 18–27 wt%, according to Japanese regulations. Prior to the application of the coating, the inner surface of the bottom plates were sandblasted to a surface finish of type ISO-Sa 2-1/2. The primer was applied using a brush or roller, and then the middle and top coats were applied by an airless spray. Each layer was fully cured before application of the next layer.

Table 2 shows the periodic inspection results for the tank. Since defects were repaired at every inspection to comply with the regulation, the "percentage of the defect area" column indicates new defects (flaws and blisters) detected at each inspection; that is, they are not cumulative. As shown in Table 2, the percentage of the defect area detected by visual observation indicates almost no defect occurrences during each oil storage period. Although the coating maintains almost complete integrity according to visual indicators, it is likely that undetected degradation has been progressing, because the coating has been in use for a long time.

Table 2 Periodical inspection results of the coating on the internal bottom plate of the oil storage tank.

| Inspection number | Period in oil (Months) | Percentage of the defect area by visual observation[a] (%) |
|---|---|---|
| 1 | 37 | 0.02 |
| 2 | 97 | 0 |
| 3 | 171 | 0.02 |
| 4 | 266 | 0.04 |

[a] Percentage of defects are detected at each inspection. The defects include flaws and blisters.

3.4 Impedance measurements and analysis

3.4.1 Laboratory experiments

Fig. 3(a) shows the experimental setup for a conventional two-electrode impedance measurement using an LCR meter (4284A, Agilent Technologies, Inc.). One side of the measurement cable of the LCR meter was connected to a platinum counter-electrode, and the other to the steel plate of a coating test piece. A PC with dedicated control software was connected to the LCR meter via a general-purpose interface bus cable. The impedance spectra were obtained over a frequency range of 20–10,000 Hz with a 1-$V_{p\text{-}p}$ sinusoidal perturbation (9 experimental points per each spectrum). Although this perturbation was larger than that of typical impedance measurements (e.g., 10–20 mV) [12–14], almost no polarization of the coating/steel interface occurred because it could be assumed that the impedance of the coating is much larger than that of the interface. In addition, we confirmed that the impedance spectrum of the non-defective coatings showed almost no variation against the change in amplitude from 10 mV to 1 V. Therefore, a sinusoidal perturbation of 1 $V_{p\text{-}p}$ was adopted to enhance the signal-to-noise ratio. The average ambient temperature was approximately 290 K. The impedance spectra were analyzed using the Zview software (Scribner Associates, Inc.).

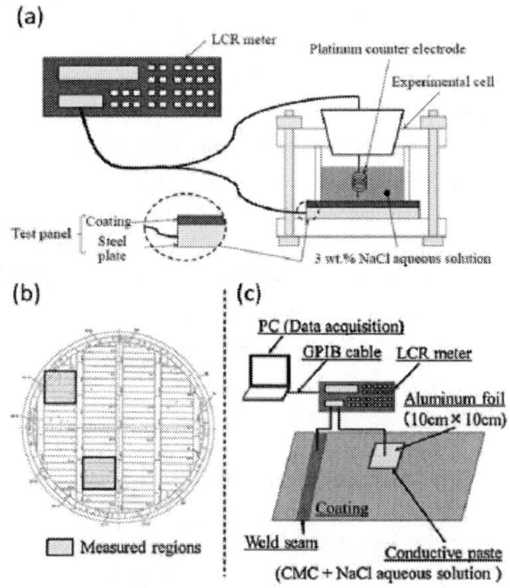


Fig. 3 Schematic diagram of impedance measurements using LCR meter for laboratory tests (a), the measured regions on internal bottom plate of the oil storage tank (b), and setup for field measurements (c).

3.4.2 Field surveys

Fig. 3(b) shows the top view of the internal bottom plates of the oil storage tank. The shaded regions in the figure represent the 10 steel plates where impedance was measured. We obtained 4 measurements per plate (40 data points in total). The data reproducibility was confirmed by at least two measurements for each spectrum. None of the plates from which field measurements were taken exhibited visual defects of the coating. The average coating film thickness from 40 measuring points was 922 μm with a standard deviation of 112 μm. Fig. 3(c) shows a schematic diagram of the two-electrode impedance measurement setup. Conductive paste composed of an aqueous solution of 5 wt% carboxyl methyl cellulose and 3 wt% NaCl was applied to the coatings as an electrolyte. Then, aluminum foil was fixed on the paste as a counter electrode (the electrode area was 100 $cm^2$). The excess of conductive paste that strayed from the aluminum foil was carefully wiped off. One side of the measurement cable of the LCR meter was connected to this aluminum foil, and the other to the steel plate where the coating was removed for weld seam inspection. The other experimental conditions were identical to those described in Section 3.4.1.

## 4. Results and discussion

### 4.1 Degradation behavior of coating systems in accelerated aging tests

Visual appearances of the coating test specimens after aging for 105 days are summarized in Fig. 4. Except for the coating system consisting of 50 μm of primer and 900 μm of glass flake coating, all specimens had blisters, some of which covered the whole surface area of the specimen. The adhesion test results showed that all blistered specimens showed separation at the interface between the primer coating and steel plate. The strengths of the blistering coating ranged from 1.5 to 2.0 MPa. Compared with the initial strengths of about 5 MPa, it is obvious that adhesion of the coating to the steel degraded considerably.

Fig. 5 shows the appearance of the steel surfaces after the adhesion tests. Black corrosion spots were observed, many of which had diameters that almost corresponded to the blister diameters. X-ray diffraction measurement identified that the black corrosion was composed of magnetite and hematite.

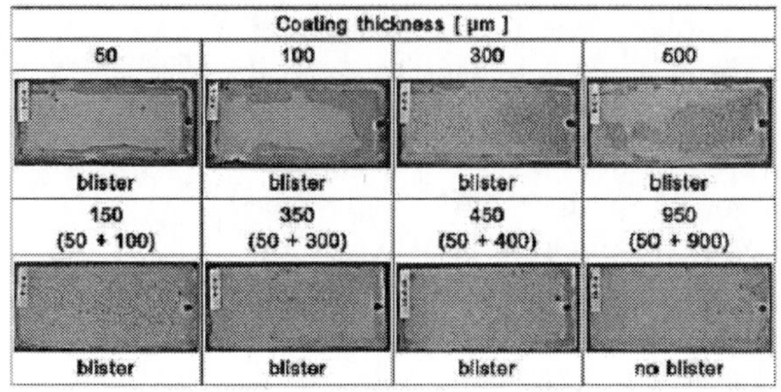


Fig. 4 Visual appearance of coating surfaces after aging tests.

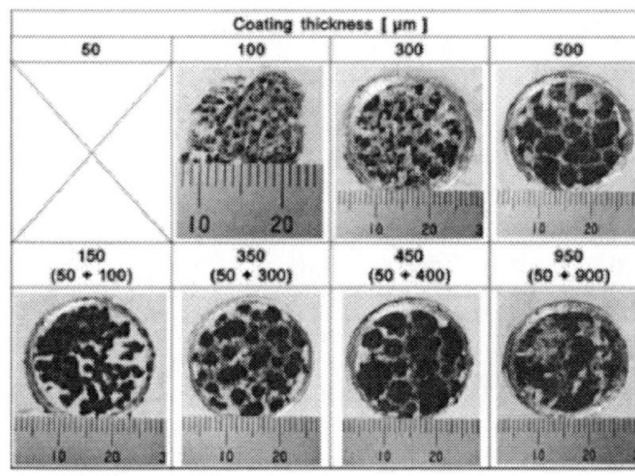


Fig. 5 Visual appearances of steel substrate surfaces after adhesion tests.

### 4.2 Electrochemical impedance characteristics of coating systems

#### 4.2.1 Results of laboratory experiments

Fig. 6 shows the change in impedance behavior at 0.1 kHz over time. As expected, the thicker coatings corresponded to larger absolute values of impedance for both coating systems. Sharp decreases in the impedances of the primer coating were observed after the initial stage of immersion. On the other hand, the values for the glass flake coating remained almost constant or slightly decreased with time. This indicates that the barrier property of glass flake coating is superior to that of primer coating. Although considerable blistering was observed in this aging test, absolute values of impedance stayed at a relatively high level near -90°. This feature of phase shift indicates the characteristics of an ideal insulator. Not only the barrier property of the coating but also the adhesion strength between the coating and steel showed accelerated degradation during this aging test. In the case of the 50-μm primer-only coating, the phase shift increased from -88 to -75° in the initial stage of immersion and finally reached -45°. The phase shift for the 150-μm-thick glass flake coating suddenly and significantly changed at 67 days. It is surmised that both effects could be caused by a localized defect, such as a pinhole, that allowed the aggressive test solution to easily penetrate the coating surface and reach the coating/steel interface. On the other hand, gradual changes in impedance behavior until these breakdown points would result from degradation of the coating barrier property.

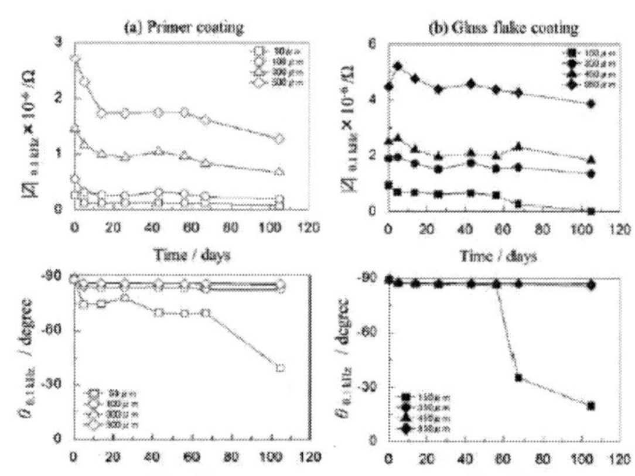


Fig. 6 Changes in the impedance modulus and the phase angle at 0.1 kHz obtained for (a) primer coating and (b) glass flake coating.

To precisely quantify this type of degradation, the change in impedance behavior was monitored by measuring impedance using a high-precision meter for 85 days of immersion in a 3 wt% NaCl aqueous solution. Fig. 7 shows the change in phase shift over time for a vinyl ester resin organic coating with specifications almost identical to those of actual coating systems applied to inner bottom plates of oil storage tank. (In this experiment, the resin containing glass flake were over-coated three times on 50 μm of primer coating, for a total thickness of 466 μm.)

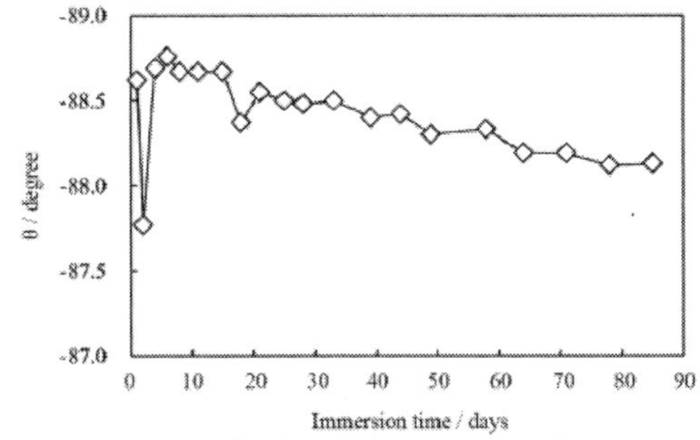


Fig. 7 Change in the phase angle at 1 kHz after immersion in 3 wt.% NaCl aqueous solution.

Although the change was very small even after 85 days, a gradual increase of the phase shift with immersion time could be detected. This change could be directly related to water uptake in the polymer matrix. To analyze this process in more detail, we obtained the impedance spectra of this coating system over a frequency range from 1 to 10 kHz, as shown in Fig. 8. Although a straight line almost perpendicular to the real axis (*Z'*) of the complex plane roughly estimates that the coating system behaves as a nearly ideal dielectric film, a slight deviation from dielectric property was clearly captured as a signal of degradation. Furthermore, this deviation, which corresponded to the phase shift, was found to be independent of frequency, as shown in the Bode plot. These impedance spectra characteristics could be well-fitted by using CPE as an equivalent circuit model. Therefore, it is suggested that the parameters of CPE would be effective indexes for residual service life estimation of coating systems.

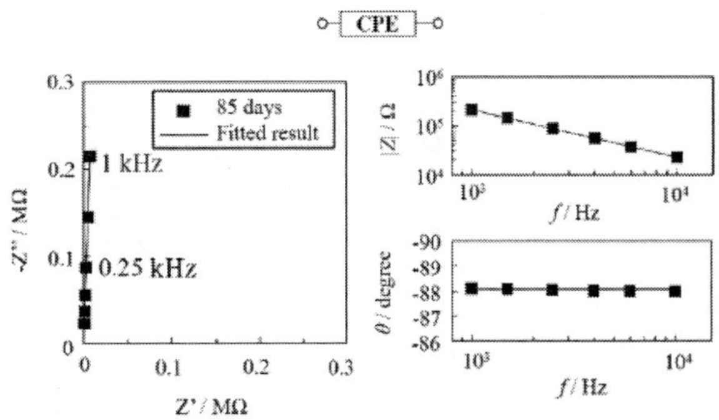


Fig. 8 Impedance behavior for a vinyl ester resin organic coating containing glass flakes coating applied on a steel plate after 85 days immersion in a 3 wt.% NaCl aqueous solution and the fitted result using one CPE model

4.2.2 Results of field surveys and the practical coating degradation model

As a next step, we surveyed the impedance behavior of actual coating systems applied to the inner bottom plates of a national oil storage tank that had undergone long service periods. All impedance spectra (40 field data points) obtained for a vinyl ester resin organic coating containing glass flakes applied on the internal bottom plate of the oil storage tank are shown in Fig. 9 as Nyquist and Bode plots. The impedance characteristics showed considerable deviation from the ideal dielectric behavior of the coating. Furthermore, there were some impedance characteristics that could not be expressed by using only one CPE. Since the coating was exposed to the electrolyte for over 22 years, it is supposed that non-uniform electrolyte solution uptake in the coating makes the impedance spectra considerably complex. Then, we used three types (A, B, and C) of equivalent electrical circuit (EEC) models to evaluate impedance behavior of the degraded coating, as shown in Fig. 10. Van Westing et al. [16] suggested that the degradation model for a high-performance anticorrosion organic coating could

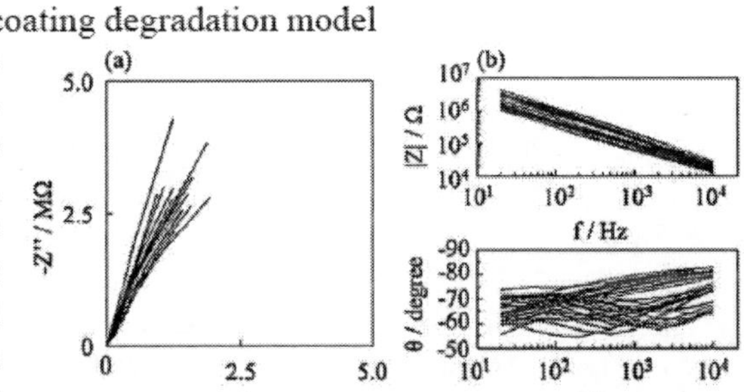


Fig. 9 All impedance spectra obtained for the coating on the internal bottom plate of the oil storage tank after 266 months exposure in the corrosive electrolyte: (a) Nyquist and (b) Bode plots.

be explained by the parallel combination of two CPEs. The model is based on the assumption that uptake of water and ions in the coating is not homogenous and the extent of uptake is different at each location due to the long "oil-in" exposure period. This dispersion could be roughly divided into two representative characteristics, which correspond to two CPEs. One indicates the area where aggressive substances easily penetrate into the coating and cause a relatively fast degradation rate. The other expresses an impedance characteristic of the area for which the barrier property remains in good condition. The impedance spectra of Type A were analyzed with a parallel combination of two CPEs, as shown in Fig. 11. The fitted curves analyzed with this model are noted in the figure as dotted lines. These curves are in good agreement with the field data. The goodness-of-fit parameter $\chi^2$ (Chi-square) [17] was calculated by Zview software. The average value of the 13 spectra of $\chi^2$ fitted for the EEC model of two CPEs in parallel was calculated to be approximately $2.4 \times 10^{-4}$, whereas $\chi^2$ using one CPE is approximately $3.4 \times 10^{-2}$. The analysis of $\chi^2$ by the parallel combination of the CPE and the resistance corresponding to conductive pathways [18] yielded a value of approximately $1.1 \times 10^{-2}$. These results indicate that the EEC of two CPEs in parallel describes the coating degradation characteristics accurately. The typical CPE parameters obtained for the fitted curves presented in Fig. 11 are summarized in Table 3. The $n$ values of $CPE_2$ show considerable deviation from the ideal dielectric properties ($n = 1$) in comparison with that of $CPE_1$. In addition, the $T$ values, corresponding to the electrical conductivity of $CPE_2$, are much larger than those of $CPE_1$. Some researchers [19,20] have already proposed a similar approach in which a coating can be separated into two different conductive regions, called "**D**" and "**I**" types. It was determined that the **D** type area has lower crosslinking and electrical resistance than the **I** type area. According to Mayne and Scantlebury [19], the **I** type area had a conductivity of $1 \times 10^{-8}$ to $1 \times 10^{-12}$ $\Omega^{-1}$, while the conductivity of **D** type area was in the range of $1 \times 10^{-5}$ to $1 \times 10^{-8}$ $\Omega^{-1}$. By comparing the reference values to our fitted results in Table 3, it is found that the $T$ values of $CPE_1$ and $CPE_2$ are in the range of the conductive values of the **I** and **D** types, respectively. Based on these results, we conclude that the coating degradation model can be described by two distinctive parts in the horizontal direction of the coating. In this study, one part is defined as the intact coating ($CPE_i$) that shows very little water uptake or degradation. The other is defined as the deteriorating coating ($CPE_d$), where easier water uptake and degradation is occurring. On the other hand, there are still some impedance spectra that could not be explained, even using the EEC model with two CPEs. It is supposed that these might indicate where degradation of the coating is progressing more

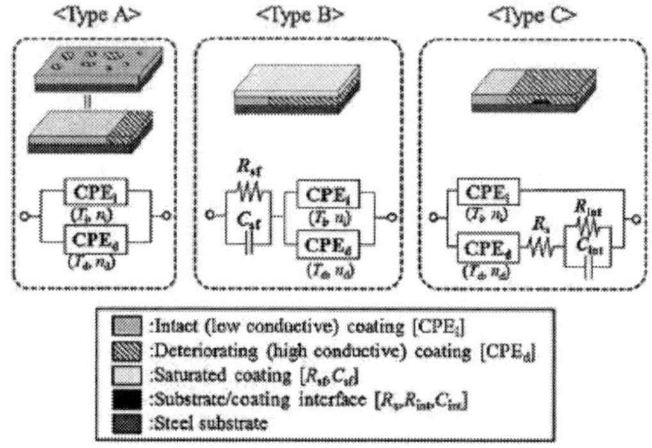


Fig. 10 Degradation models of the coating on the internal bottom plate of the oil storage tank after a long immersion duration.

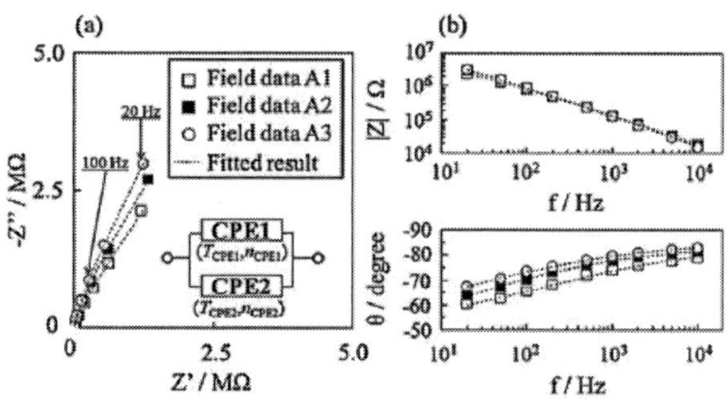


Fig. 11 Typical impedance plots (Type A) obtained for the coating on the internal bottom plate of the oil storage tank after 266 months exposure in the corrosive electrolyte: (a) Nyquist and (b) Bode plots.

Table 3 Fitted value for parameters in the equivalent circuit model consisting of a parallel combination of two CPEs.

| Field data | $T_{CPE1} \times 10^9$ ($s^n$ $\Omega^{-1}$) | $n_{CPE1}$ (–) | $T_{CPE2} \times 10^9$ ($s^n$ $\Omega^{-1}$) | $n_{CPE2}$ (–) |
|---|---|---|---|---|
| A1 | 1.38 | 0.943 | 20.9 | 0.545 |
| A2 | 1.85 | 0.934 | 16.3 | 0.501 |
| A3 | 1.58 | 0.954 | 11.4 | 0.555 |

severely, although no defects were seen in the visual observation. Consequently, the additional time constant factors appeared and had a complicated influence on the impedance spectra. Recently, some researchers [21–23] suggested that the coatings are divided into two layers, with water permeation characteristics varying according to depth. It has been recognized that the outer layer in contact with the solution is saturated by water and the inner layer near the substrate is regarded as almost dry [23]. By considering the immersion duration of this coating, it is possible that some of the saturated layer forms near the solution side of the coating. Therefore, the degradation model could be explained by the parallel combination of a saturated capacitance $C_{sf}$ and resistance $R_{sf}$ in series with the two CPEs in parallel, as shown in Fig. 12 (Type B).

As discussed in the previous section, extra water uptake could occur in the deteriorating coating ($CPE_d$). Therefore, it is supposed that much absorbed water reaches the substrate through the $CPE_d$ layer. Thus, the degradation model could be described with the series combination of $CPE_d$ and the interfacial impedance ($R_s$, solution resistance of the substrate/coating interface; $R_{int}$, interfacial protection resistance; and $C_{int}$, interfacial protection capacitance) in parallel with the $CPE_i$, as shown in Fig. 13 (Type C). When water reaches the substrate, anodic and cathodic sites are established at the coating/substrate interface (micro-cell corrosion). By anodic dissolution and deposition of the corrosion products resulting from the corrosion reactions, a rust layer is formed at the interface [24,25]. These corrosion products continue to accumulate within the layer, which becomes more dense. Eventually, the supply of oxygen, which causes a cathodic reaction at the cathodic sites, is diffusion-limited by the rust layer and the corrosion reactions are inhibited. It is implied from this process that the interfacial impedance is high enough to have an effect on the impedance of the coating itself. Further investigation is required to clarify the interfacial degradation mode.

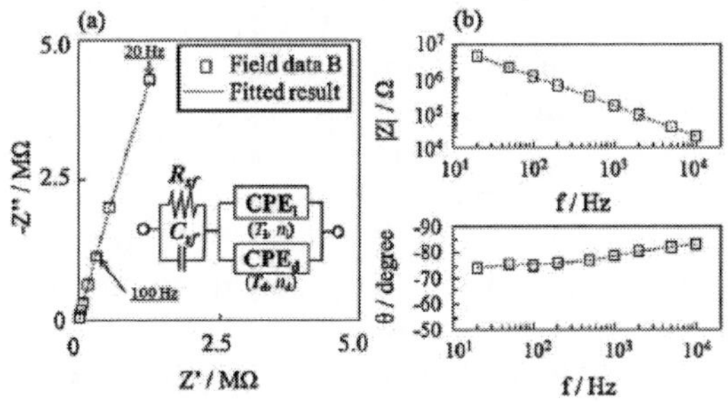


Fig. 12 Impedance plots (Type C) obtained for the coating on the internal bottom plate of the oil storage tank after 266 months exposure in the corrosive electrolyte: (a) Nyquist and (b) Bode plots.

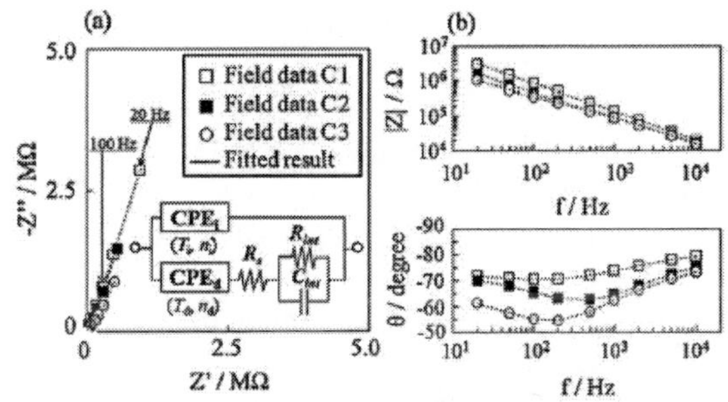


Fig. 13 Typical impedance plots (Type C) obtained for the coating on the internal bottom plate of the oil storage tank after 266 months exposure in the corrosive electrolyte: (a) Nyquist and (b) Bode plots.

Based on the degradation models, the Type B and C impedance plots were analyzed according to the following procedure. First, assuming that water does not reach the substrate, the analysis was conducted on the Type B model. Among the field data, only one spectrum was fitted to this model, as shown in Fig. 12. The second spectrum, for which it is supposed that the water reaches the substrate and the interfacial impedance has an influence on the spectra, the experimental plots were fitted on the Type C model. Fig. 13 shows typical impedance plots and fitted curves. The remaining 26 spectra showed good fitting results. Table 4 shows the CPE parameter values obtained from fitting curves in Figs. 12 and 13. As can be seen in Table 4, the CPE parameter values for Type C showed less capacitive behavior (lower $n$ values and higher $T$ values) than that of Type B. The result coincides with the consideration that the water uptake of the coating in the Type C model progressed considerably. (In other words, the interfacial impedance was enough to

Table 4 Comparison between fitted parameter values in different equivalent circuit models of Type B and C.

| Field data | $T_i \times 10^9$ ($s^n\,\Omega^{-1}$) | $n_i$ (-) | $T_d \times 10^9$ ($s^n\,\Omega^{-1}$) | $n_d$ (-) |
|---|---|---|---|---|
| B | 1.01 | 0.941 | 8.73 | 0.592 |
| C1 | 2.30 | 0.906 | 21.8 | 0.526 |
| C2 | 3.57 | 0.883 | 51.4 | 0.570 |
| C3 | 5.37 | 0.849 | 136 | 0.469 |

affect the coating impedance itself, as mentioned previously.) Thus, 27 field data points of these impedance spectra can be analyzed by using the EEC corresponding to each degradation model.

4.4 Predictive diagnosis based on an extreme value distribution

It was found that the degradation of coating at a specific measuring point could be quantitatively evaluated using CPE parameter values extracted from EEC analysis. However, an estimation of residual life of the coating system as a whole is still technically challenging because measurements vary significantly from point to point. Environmental conditions, such as temperature, chemical composition, and mechanical stress on the vast expanse of tank bottom plates, are not constant and easily vary from hour to hour, depending on the meteorological and even geological phenomena at the tank yard. Thus, degradation behavior tends to be very localized. For these reasons, there is a risk of overestimating residual service life if the evaluation method is based on averaged data, which could lead to serious corrosion accidents.

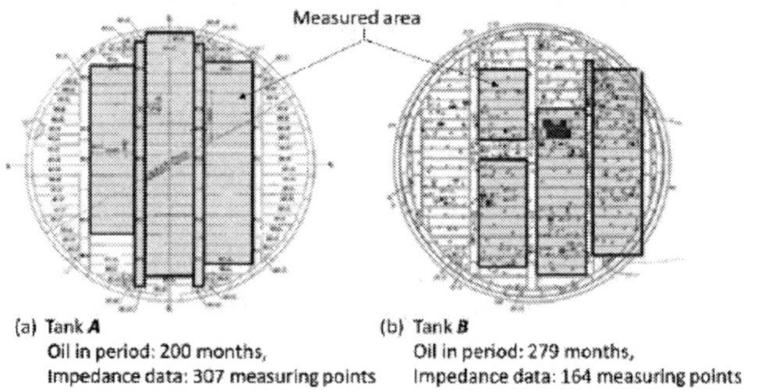


Fig. 14 Field impedance measurements for actual oil storage tanks.

In this study, the Gumbel distribution [26], an extreme value distribution, was used for statistical analysis of CPE parameter values and to establish a reasonable diagnosis method. Recently, this distribution was used to statistically predict the maximum depth of pitting corrosion [27]. The cumulative distribution function of the Gumbel distribution is expressed as the following equation:

$$F(x)=\exp\left[-\exp\left(-\frac{x-\lambda_1}{\sigma_1}\right)\right], \quad -(3)$$

where $x$ is the observed extreme value $(1 - n_d)$ and $\lambda_1$ and $\sigma_1$ are constants that determine the shape of the distribution.

In order to demonstrate the applicability of the Gumbel distribution to statistical analysis of CPE parameter values obtained from an actual oil storage tank, field impedances were measured in two tanks (capacity of 110,000 kL, inner diameter of 81.5 m) with inner bottom plates coated with nearly identical vinyl ester resin coatings and exposed to similar environmental conditions, as shown in Fig. 14. The coating in the measured area had no visible defects and impedance data were analyzed by the EEC model with two CPEs. Values for the determined CPE parameter $n_d$ of the two tanks were 307 points for Tank A and 164 points for Tank B. Extreme values were extracted from the data set, which consisted of 8-11 randomly selected samples.

The Gumbel plots of maximum values of CPE parameter $n_d$ obtained from field measurements for tanks A and B are given in Fig. 15. An almost linear relationship was obtained for both tanks, indicating that the maximum values based on this CPE parameter could be treated by the theory of the extreme value statistics. The value of $\lambda_1$ for Tank A was larger than that for Tank B, indicating that the degradation of the coating system was more severe than that of Tank A, since the value of $x$ corresponds to the degree of

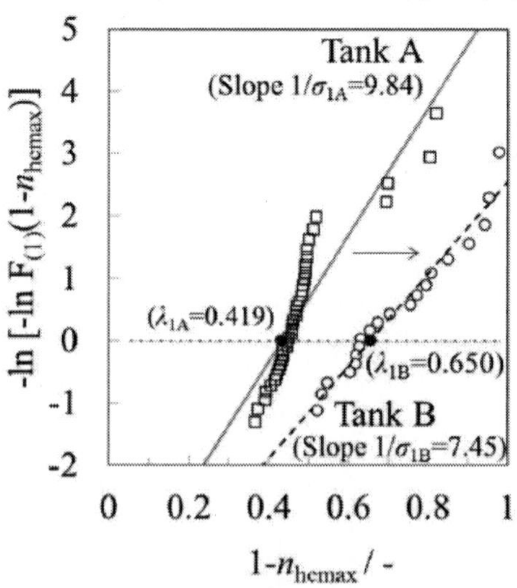


Fig. 15 Gumbel plots using impedance parameter obtained from actual oil-storage tanks.

degradation. This result was consistent with the tank profiles with respect to the oil-in periods. Furthermore, the values of $\sigma_1$, which corresponds to the slope of the Gumbel plot, tend to become smaller as the oil-in period increases. It is suggested that the shape of the extreme value distribution becomes broader and the risk of severe localized degradation increases.

## 5. Conclusion

This study undertook the development of a diagnostic method based on electrical impedance analysis for evaluating residual service life of vinyl ester resin coatings containing glass flakes. As a result of laboratory experiments and tracking of various degradation behaviors of coatings applied to actual oil storage tanks, it was found that their characteristics could be expressed using EEC models with CPEs. Although many issues still need to be clarified, the EEC model proposed in Fig. 16 seems to be plausible for heavy-duty coating systems. Among the elements, $CPE_d$ would be the most important index to evaluate degradation and failure of coating systems. The extracted parameters of $CPE_d$ were analyzed using extreme value statistics as a tool for service life assessment. It was found that the extreme value distribution of the $CPE_d$ parameter followed the Gumbel statistical model. Future work will clarify and verify the dependence of shape parameters on oil-in periods, and we expect to realize a practical diagnosis of residual service life, as shown by the concept in Fig. 17.

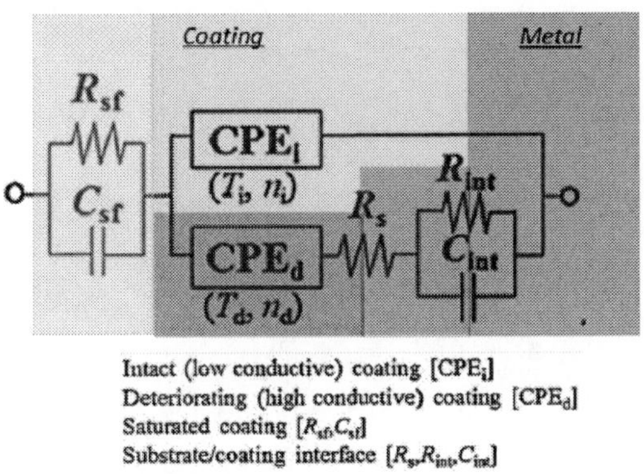


Fig. 16 Electrical equivalent circuit model for coting degradation.

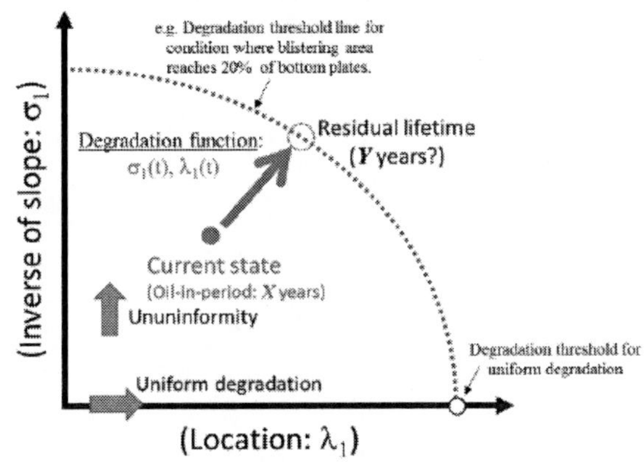


Fig.17 How to predict a residual lifetime of the heavy-duty coating by the application of extreme value statistics for the electrical properties of the coatings.

## Acknowledgment

This research was performed with financial support from the Japan Oil, Gas and Metals National Corporation.

## References

[1] N. Sridhar, C.S. Brossia, D.S. Dunn, A. Anderko, Corrosion (NACE) 60 (10) (2004) 915.
[2] R.E. Melchers, Prob. Eng. Mech. 23 (2008) 482.
[3] J.F. Bates, Mater.Performance 8 (1) (1969) 33.
[4] A. Miszczyk, K. Darowicki, Corros. Sci. 40 (1998) 663.
[5] M. Ehsani, H.A. Khonakdar, A. Ghadami, Prog. Org. Coat. 76 (2013) 238.
[6] W. Funke, Prog. Org. Coat. 9(1981) 29.
[7] O. Schneider, R.G. Kelly, Corros. Sci. 49 (2007)594.
[8] M. Itagaki, A. Ono, K. Watanabe, H. Katayama, K. Noda, Corros. Sci. 48 (2006)3802.
[9] S. Scale, V. Dolecek, M. Slemnik, Prog.Org. Coat. 62 (4) (2008) 387.
[10] W. Tian, L. Liu, F. Meng, Y. Liu, Y. Li, F. Wang, Corros. Sci. 86 (2014) 81.
[11] B. Hirschorn, M.E. Orazem, B. Tribollet, V. Vivier, I. Frateur, M. Musiani, J. Electrochem. Soc. 157 (12) (2010) C452.
[12] C.A. Schiller, W. Strunz, Electrochim. Acta 46 (2001) 3619.

[13] J.-B. Jorcin, M.E. Orazem, N. Pebere, B. Tribollet, Electrochim. Acta 51 (2006) 1473.
[14] S. Amand, M. Musiani, M.E. Orazem, N. Pebere, B. Tribollet, V. Vivier, Electrochim. Acta 87 (2013) 693.
[15] E.P.M. Van Westing, G.M. Ferrari, Prog.Org. Coat. 23 (1993) 89.
[16] E.P.M. Van Westing, G.M. Ferrari, J.H.W. De Wit, Corros. Sci. 36 (6) (1994) 979.
[17] B.A. Boukamp, Solid State Ionics 20 (1986) 31.
[18] S. Skale, V. Dolecek, M. Slemnik, Corros. Sci. 49 (2007) 1045.
[19] J.E.O. Mayne, J.D. Scantlebry, Br. Polym. J. 3 (5) (1971) 237.
[20] E.O. Eltai, J.D. Scantlebury, E.V. Koroleva, Prog. Org. Coat. 73(2012) 8.
[21] J. Kittel, N. Celati, M. Keddam, H. Takenouti, Prog. Org.Coat. 41 (2001) 93.
[22] B.R. Hinderliter, S.G. Croll, D.E. Tallman, Q. Su, G.P. Bierwagen, Electrochim. Acta 51 (2006) 4505.
[23] G. Bouvet, D.D. Nguyen, S. Mallarino, S. Touzain, Prog. Org. Coat. 77 (12A)(2014) 2045.
[24] J. Kittel, N. Celati, M. Keddam, H. Takenouti, Prog. Org. Coat. 46 (2003)135.
[25] C. Lin, T. Nguyen, M.E. McKnight, Prog. Org. Coat. 20 (1992) 169.
[26] E. J. Gumbel, Statistics of extremes, Colombia Univ. Press (1958).
[27] P. M. Aziz, Corrosion, 12 (1956) 495.

Paper No. 5

# Applications of process and kinetic models to optimize the operation of Hydrocracker

***Rashid M. Ansari***
*PhD (Curtin), MEngSc (Adelaide), MPH (Qld), MClinEpi (Newcastle)*
*Engineering Specialist – Refining development, R&DC*

**Abstract**

**Objectives:** The main objective of applications of process and kinetic models in developing the Real-Time Optimization (RTO) strategies for hydrocracking process was to help refinery operation to maximize contribution of the plant to the business profit, providing best in class performance, optimizing the plant operation, enhancing safety and reliability.

**Methods:** The process and kinetic models were developed and incorporated in real-time optimization framework and successfully implemented on hydrocracking process to enhance the yield of high value products. The dynamic model of the process was developed using plant identification techniques and integrated with real-time optimization applications. Finally, a closed-loop quality control system was established to keep the product specifications on targets.

**Results:** The applications of process and kinetic models within the framework of Real-Time Optimization helped to optimize key process operating variables by shifting the unit margin towards the optimum and operations were better placed to challenge targets and operating conditions, driving the plant toward a more profitable operating regime, bringing the higher benefits. The application of process and kinetic models resulted in improving the yield of diesel and gasoline and increasing the feed rate subject to unit constraints and catalyst run length. In addition, the use of process and kinetic models together with data reconciliation system adjusted the reactors bed inlet temperature to maintain a desired WABT and protected against constraints such as hydrogen quench and delta temperature limits and improved the catalyst life by maintaining the specified reactor bed temperature profile.

**Conclusion:** The applications of process and kinetic models within the framework of real-time optimization brought substantial benefits to the refinery. Of particular mention are the benefits in improvement of the basic data for planning, improvement in process understanding and monitoring, providing economic focus on unit operation, allowing the use of process and kinetic models for off-line and on-line studies, active constraint tracking and pushing and also the monitoring of heat exchangers fouling and compressor efficiency in real-time. It is envisaged that these applications will be extended to other processing units to optimize the overall operation of the refinery.

**Key words:** Real-time optimization; model-predictive control; Hydrocracker; refinery processes; economic objective function

## Introduction

The combination of process and kinetic models are well known to improve the performance in real-time operation of the processing units. These models were developed and incorporated in real-time optimization framework and successfully implemented on hydrocracking process to enhance the yield of high value products. The Real-Time Optimization (RTO) is the process of finding the set of conditions required to obtain the best economic result for a given condition, a method of determining a set of operating targets for the online control system to achieve the maximum profit possible within constraints and practical limits of the process equipment while keeping all the products at specifications [1, 2, 3]. RTO systems are a combined set of techniques and algorithms that continuously evaluate process operating conditions and implement business-focused decisions to improve process performance [4, 5].

The two-step approach, a model-based technique, is the most common (and possibly the only) static real-time optimization strategy available in commercial RTO systems [6, 7, 8]. Its name derives from the procedure employed for determining the set of decision variables, where plant information is used to update model parameters based on the best fitting of measurements in the first step, and afterwards, the updated model is used to calculate the set of decision variable values that are assumed to lead the process to its best economic performance. RTO systems are widely used in the petroleum refineries as part of modern day control systems [9, 10], but may also be found in other sectors of the industry [11].

The advantages of RTO are attributed to the use of a priori information in the form of a process model, and model-based techniques may present superior performance among others. In general, the more accurate the model is, the better will be the RTO system performance [12, 13]. Therefore, RTO applications are typically based on rigorous steady-state models of processes. It has long been shown that manipulation of model parameters to fit available process measurements does not necessarily guarantee the construction of an adequate model for process optimization [14, 15].

This article aimed at presenting the implementation techniques and performance evaluation of RTO systems on a hydrocracking unit on a model case refinery in the Middle East. The aim is to highlight some features of RTO systems incorporating the process and kinetic models that have brought higher benefits to the refinery, by maximizing the economic objective functions and enhancing the yield of diesel and gasoline products.

## RTO System Integration

The basic instrumentation and control layer is established for the plant by installing the Distributed Control System (DCS) so that all the specifications of the units are met and operation is stabilized and run safely and smoothly. The next step is to implement advanced process control technologies on the plant. That control layer takes care of process constraints, interactions and future consequences of the current actions in the operation of the process units.

This control layer is called model predictive control and inferential models, which form a closed loop quality control system on the unit. The next control layer is the Real-time optimization which provides global view and incorporate different aspects and interrelations among processes in a model to compute operational decisions that optimize process efficiency and economy.

Figure 1 shows various layers of automation integrating with Real-Time Optimization. The Refinery LP model will set the target for the optimizer, which will use the steady state and kinetic models to provide the optimized set points for Multivariable controller (MVC), to implement with the processing units, using the Distributed Control System (DCS) as a platform. The inferential models interfacing with MVC form a closed-loop quality control system.

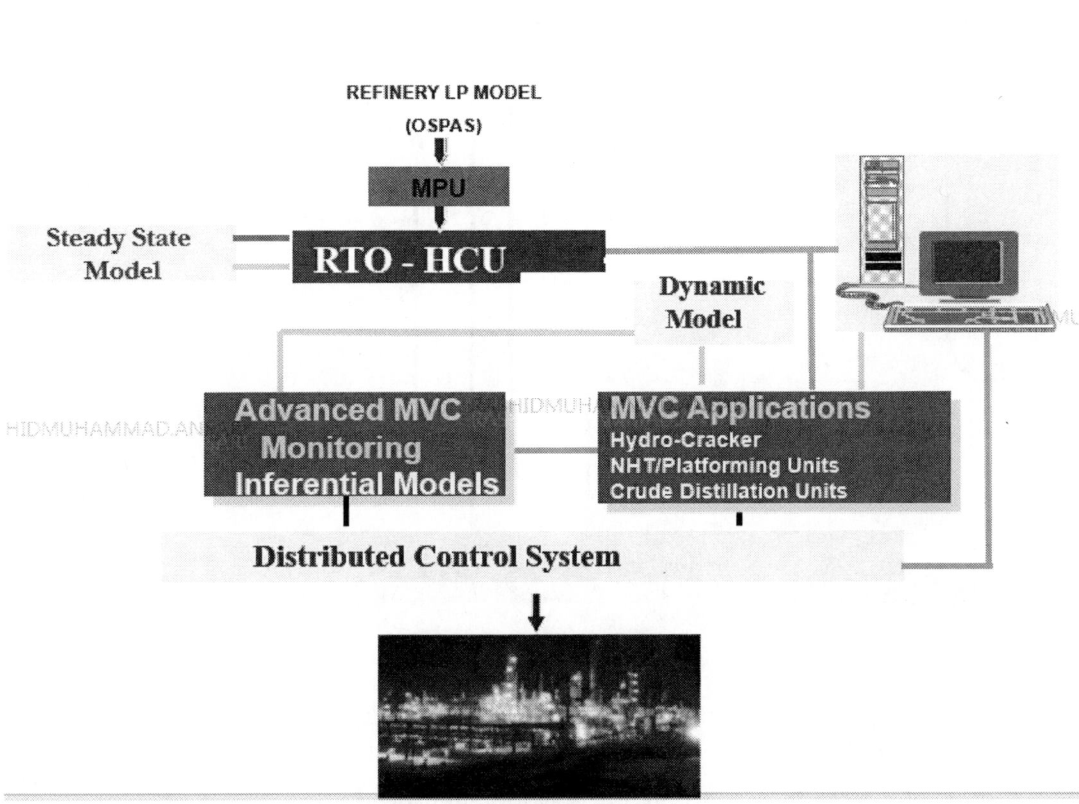


**Figure 1:** Real-Time Optimization System Integration

## Integration of Inferential Models

The inferential models or Robust Quality Estimator (RQE) uses selected measurements (such as feed rate, temperature, pressure, etc.) to predict the real value of a critical product property or complex process variable, that cannot be measured online or, is measured infrequently with potential large time delay. In a typical application, RQE provides a continuous real-time calculated value for closed loop control. It can cope with nonlinear systems, varying gains and uncertain process dynamics. This unique feature makes the model-predictive controller robust to varying process conditions and nonlinearities.

Figure 2 gives a simplified example of the systematic approach to construct the naphtha FBP inferential model with process variables, such as pressure-compensated temperature, pressure of the column and reflux ratios. The model is updated using laboratory results. A system of inferential model update is shown in the lower section of Figure 2, providing laboratory result as a basis to update the predicted values and a first-order filter to compensate for the noise in the flow measurements. The average deviation from the laboratory results is only 2 °C, and considering the repeatability of ± 4 °C in Naphtha FBP laboratory results, this prediction can be considered very accurate.

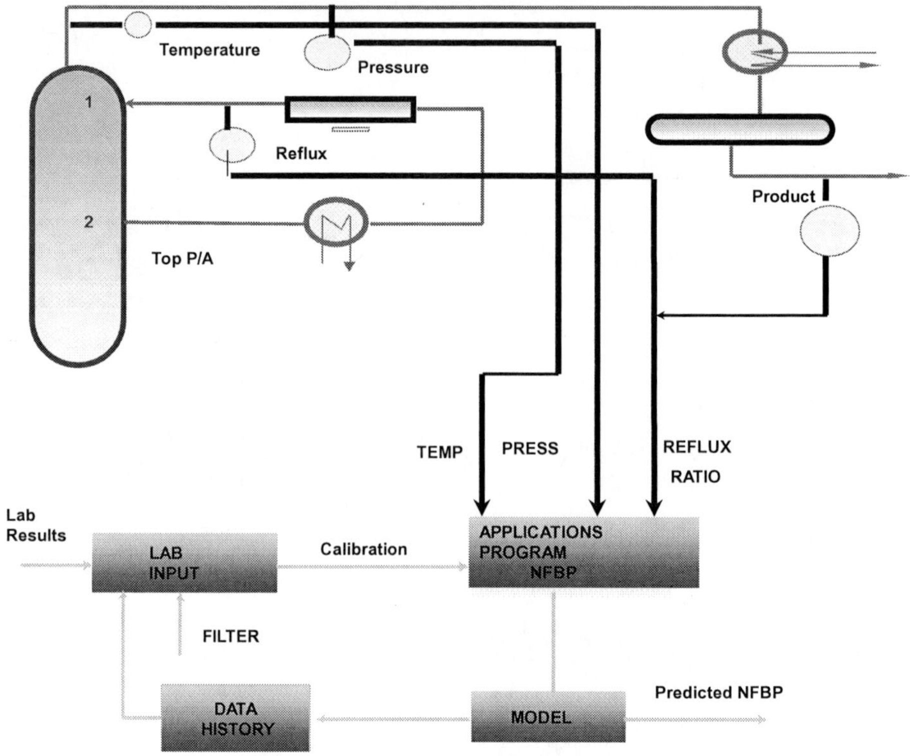


**Figure 2** System of inferential model and its development procedure

## Process Description

The Hydrocracker Unit (HCU) of the model case refinery has maximum feed capacity of 27,500 – 28,500 barrel per stream day. The HCU consists of three major sections: Reactor Section, Debutanizer, and Main Column. The Hydrocracker Unit operation objective is to maintain the feed flow at the required rate, while maximizing the gasoline or diesel yield, subject to the products qualities of the main fractionator. The key focus of RTO implementation was to provide a combination of operating conditions, which will support an increase in each reactor temperature and maximize the economic objective function to enhance the yield of gasoline or diesel products. Therefore, raising the first stage (DHC) reactor temperature will enhance the diesel yield, while raising the second stage (HC) reactor temperature will increase the gasoline yield.

## Current Operating Strategy

The current HCU operating strategy at model case refinery can be summarized as follows:

- Maximize the unit throughput, whilst respecting unit constraints and required catalyst run length.
- Adjust the stage 1 or stage 2 reactor severities, to meet the requirements of two main modes, maximum naphtha or maximum diesel.
- Optimize the circulating refluxes in the product fractionator, to reduce energy costs.

The average unit stream time is 92% (including the effects of catalyst run length), which is equivalent to 336 days per year.

### Online closed-loop optimization Steps

The following is a simplified description of the steps of operations involved in online closed-loop optimizations: the data reconciliation; rating case, base case; constraint calculation; optimization case and implementation. Figure 3 shows these steps starting from steady-state detection to implementation – downloading the targets to the control system.

### Steady-state Detection (SSD)

The steady-state detection has a great influence on RTO performance but is not discussed in details in the RTO literature. An important element of RTO systems refers to the mechanism that triggers the process optimization. Traditionally, it is based on the stationarity of measured process data and is accomplished by the SSD module. In the current RTO application, the "statistical method" of steady state detection was implemented.

### Data Reconcilliation

Data reconcilliation involves mass and heat balances around each and every piece of equipment in the unit which is fully consistent with the overall balances [16]. The benefits include consistency and accuracy of yield calculations and targeted instrument maintenance. When data present significant uncertainty, data reconciliation is the first step to be applied in a model-based approach to process optimization. The target is to estimate consistent values of all plant variables from available online measurements based on a process model. Soderstrom et al. [17] highlighted that if measurement errors resulting from poor instrument calibration are not considered, the data reconciliation step or subsequent parameter estimation step will not provide meaningful answers.

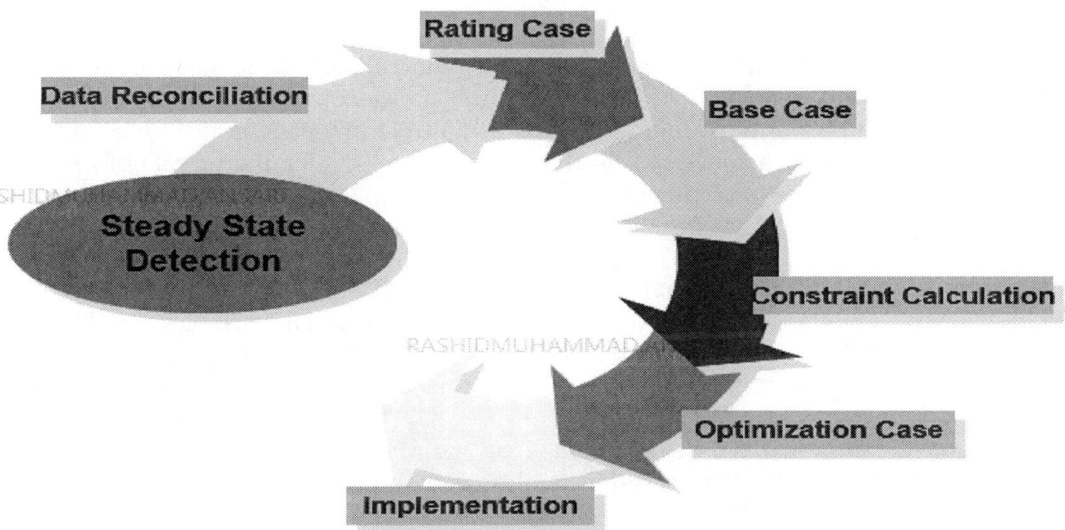


**Figure 3** Online closed loop optimization steps

The combination of RTO and regulatory control can be considered similar to cascade control system as shown in figure 4. The outer RTO loop will operate more slowly than the inner loop, and a poor design of this interaction results in poor performance. The dynamic controller handles the transformation between the steady-state model used in RTO and the actual dynamic operation of the process. If RTO model and dynamic model have very different gains, the resulting combination can perform poorly. As in cascade control, the inner loop should be faster than the outer loop; otherwise poor closed-loop performance may be resulted [18].

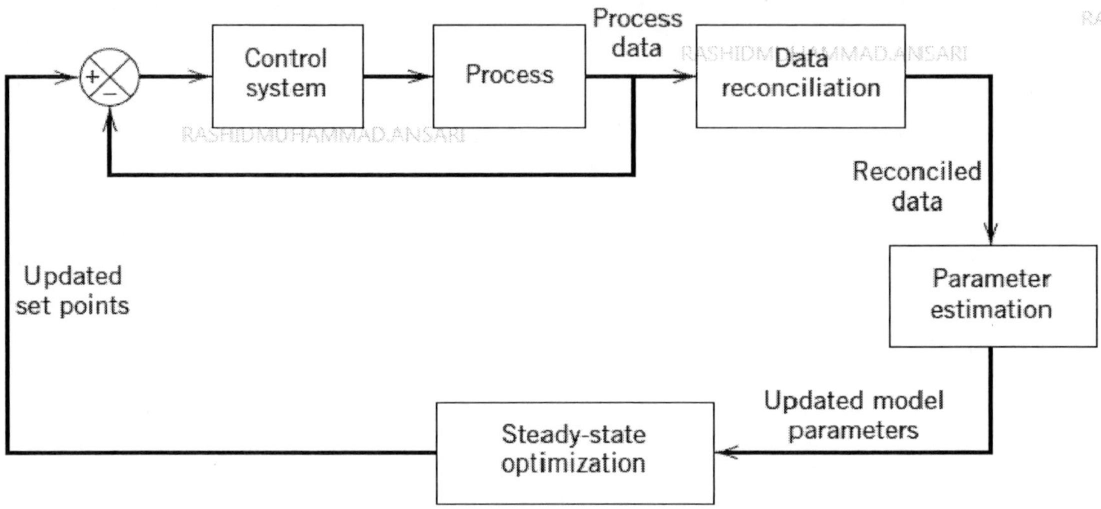


**Figure 4** A block diagram representing RTO and regulatory feedback system

In the rating case, the rating of the equipment is identified and in the base case, the economic objective function is introduced and measurement offsets are calculated. The benefits of these steps include online calculations of unit margin and additional targeted instrumentation maintenance.

The constraint calculations include equipment constraint, control constraints, process constraints and economic constraints. The benefits of this step include active constraint tracking and focused debottlenecking. In the optimization step, the effect of multiple independents is observed on an economic objective function to achieve the maximum profit and finally, the targets of optimized solutions are downloaded in the implementation step.

## Results

In relation to the benefits of RTO implementation on model case refinery in the Middle East, it was observed that benefits obtained mainly from reactors WABT independent variables. The WABT set points were implemented at each run and reached by the multivariable control applications. The feeds were limited by the constraints but contributed at the profit. The main fractionator was the most constraining factor, mainly in the LDO and HDO sections. The benefits were achieved without compromising run length and have been obtained by maintaining the Weighted Average Bed Temperature (WABT) and varying the operating conditions to improve product values.

It was one of the findings from RTO simulation data that by decreasing the Indirect Recycle Flow (IDR) from 50 $m^3/hr$ to 48 $m^3/hr$, substantial benefits can be achieved through higher unit margins, and these economic gains are maximum when HDO rundown flow is unconstrained.

The hydrocracker has the capability of running at a lower 48$m^3/hr$ IDR flow on a sustainable basis and the economic benefits for such a permanent move is expected to be significant based on RTO simulations data. This shift can be achieved with existing RTO constraints, including weekly maximum WABTs, reactor peak temperatures, and acceptable catalyst deactivation rates.

## RTO Benefit Analysis

The data used in this benefit analysis covers the RTO operation starting from March 2008 to June 2008. The data is collected from RTO online software called "Pro-Clarity." This data is based on optimized solutions, which take into account all the following four RTO benefit incentives given below, and quantifies the benefits based on every solution provided by RTO.

1. Improve the yield of high value products by maximizing the economic objective function to maximize diesel or gasoline on Hydrocracker.
2. Increase the federate subject to unit constraints and catalyst run length.
3. Adjust the bed inlet temperature to maintain a desired WABT and protects against constraints such as hydrogen quench and delta Temp limits.
4. Improve catalyst life by maintaining the reactor bed temperature profile.

RTO benefits based on the economic data obtained from Pro-clarity and by using statistical analysis was calculated by taking mean value of benefits for the 3-month run. Considering the international trends of RTO applications and utilization factor of 0.75, the net benefits of RTO applications are significant.

## Discussion

The development and implementation of RTO system on hydrocracking unit was an outcome of great team work, collaboration of industrial and vendor's teams over the past couple of years, and has brought substantial benefits to operations, specifically and overall to the refinery. It has increased the production of high value products, such as diesel and gasoline, by maximizing the economic objective functions by using Real-Time Optimization techniques. The RTO system implemented on the hydrocracker has improved overall performance of the unit and its efficiency, which was a significant step forward to further integrate other advanced control applications in the refinery.

It was also observed during implementation phase that the reliable instrumentation is the key to success of RTO. This includes the necessary support infrastructure, to ensure that any instrument faults identified during design and commissioning are quickly and effectively dealt with. It is important to have frequent and reliable inferential models (RQEs) and analyzers, and equally important is that these quality indicators continue to perform effectively.

For the optimizer, a full coverage of the modelling scope by multivariable control applications (that is dynamic modelling) is a prerequisite for a stable operation and implementable independents. Information about the actual line-up and equipment used in the plant should be available for a proper mode representation. In case the documentation is not available, the optimizer focal point will have to spend a considerable amount of extra time in the field to sort the required information. Finally, the dedication and commitment of all the team members and management was one of the main critical factors for the success of this important project.

## Conclusion

The real-time optimization of the hydrocracker was implemented successfully and brought substantial benefits to hydrocracker operation. RTO set point implementation demonstrated that siginificant financial benefits would have been easily achieved, especially with the high margin on that unit. The main benefits of RTO implementation were achieved from the reactors Weighted Average Bed Temperature (WABT) independents. These set points were implemented at each run and were reached by MVC. The feeds were limited by the constraints but contributed at the overall profit.

It is also envisaged that implementing Real-Time Optimization application on $CO_2$ emission reduction and other processing units would bring higher benefits to the refinery. Since the applications of multivariable control are already in place on the key processing units of the refinery, the implementation of RTO will help to drive the plant operation toward the optimum operation, by maximizing the economic objective functions on these processing units.

## Acknowledgment

The author wishes to express his gratitude to the refinery's RTO team members: process engineers, operation engineers, hydrocracker operators, and foremen for their participation on this project. Many thanks to the Oil Upgrading R&D Division Management for providing me with the opportunity to share the RTO implementation knowledge to R&D staff members.

## References

1. Garcia, C.E.; Morari, M. Optimal operation of integrated processing systems. Part I: Open-loop on-line optimizing control. *AIChE J.* **1981**, *27*, 960–968.
2. Ellis, J.; Kambhampati, C.; Sheng, G.; Roberts, P. Approaches to the optimizing control problem. *Int. J. Syst. Sci.* **1988**, *19*, 1969–1985.
3. Engell, S. Feedback control for optimal process operation. *J. Process Control* **2007**, *17*, 203–219.
4. Chachuat, B.; Srinivasan, B.; Bonvin, D. Adaptation strategies for real-time optimization. *Comput. Chem. Eng.* **2009**, *33*, 1557–1567.
5. François, G.; Bonvin, D. Chapter One—Measurement-Based Real-Time Optimization of Chemical Processes. In *Control and Optimisation of Process Systems*; Advances in Chemical Engineering; Pushpavanam, S., Ed.; Academic Press: New York, NY, USA, 2013; Volume 43, pp. 1–50.
6. Naysmith, M.; Douglas, P. Review of real time optimization in the chemical process industries. *Dev. Chem. Eng. Miner. Process.* **1995**, *3*, 67–87.
7. Marlin, T.E.; Hrymak, A.N. *Real-Time Operations Optimization of Continuous Processes*; AIChE Symposium Series; 1971-c2002; American Institute of Chemical Engineers: New York, NY, USA, 1997; Volume 93, pp. 156–164.
8. Trierweiler, J.O. Real-Time Optimization of Industrial Processes. In *Encyclopedia of Systems and Control*; Baillieul, J., Samad, T. Eds.; Springer: London, UK, 2014; pp. 1–11.
9. Rotava, O.; Zanin, A.C. Multivariable control and real-time optimization—An industrial practical view. *Hydrocarb. Process.* **2005**, *84*, 61–71.
10. Young, R. Petroleum refining process control and real-time optimization. *IEEE Control Syst.* **2006**, *26*, 73–83.
11. Shokri, S.; Hayati, R.; Marvast, M.A.; Ayazi, M.; Ganji, H. Real time optimization as a tool for increasing petroleum refineries profits. *Pet. Coal* **2009**, *51*, 110–114.
12. Chen, C.Y.; Joseph, B. On-line optimization using a two-phase approach: An application study. *Ind. Eng. Chem. Res.* **1987**, *26*, 1924–1930.
13. Yip, W.; Marlin, T.E. The effect of model fidelity on real-time optimization performance. *Comput. Chem. Eng.* **2004**, *28*, 267–280.
14. Roberts, P. Algorithms for integrated system optimisation and parameter estimation. *Electron. Lett.* **1978**, *14*, 196–197.
15. Forbes, J.; Marlin, T.; MacGregor, J. Model adequacy requirements for optimizing plant operations. *Comput. Chem. Eng.* **1994**, *18*, 497–510.
16. Narasimhan, S., and Jordache, C. *Data Reconciliation and Gross Error Detection*, Gulf Publishing, Houston, TX, 2000.
17. Soderstrom, TA., Edgar, TF, Russo, LP et al. Industrial Application of a Large-Scale Dynamic Data Reconciliation Strategy, *Ind. Eng. Chem. Res.* 2000, **39**, 1683
18. Marlin, T. E., and Hrymak, AN. Real-Time Operations Optimization of Continuous Processes, in *Chemical Process Control V, AIChE Symp. 1997, Ser.* **93**, No. 316, 156.

Paper No. 6

# Effect of acid medium and initial Si/Al ratio on the synthesis of mesoporous materials with enhanced acidity and hydrothermal stability

***Lianhui Ding, Hanaa Habboubi, Essam Sayed***
*Research and Development Center, Saudi Aramco*

## 1. Introduction

Hydrocracking is one of the most important heavy oil conversion processes in modern refineries and petrochemicals plants. Hydrocracking catalysts play a critical role in determining the product slates and properties, and thus the economics. Zeolites are the key cracking component of the hydrocracking catalysts, and determine the catalyst reaction performances. The pore sizes of zeolite Y and beta, widely used in preparing hydrocracking catalysts, are around 0.74nm, and too small for the large molecules in heavy oils to diffuse in and out of the internal surface where the active sites are located. To improve the mass transfer of the large molecules in heavy oils, and avoid secondary cracking, the mesoporous material with mesopores (2-50nm) and improved hydrothermal stability was studied. The effect of the acid medium and initial gel Si/Al on the properties of SBA-15 were studied.

## 2. Experimental

In-situ crystallization to form mesoporous materials was performed under different acid medium (0, 0.07, 0.5, 2.0, 5.0M HCl solution), and different Si/Al ratio (5-50). Triblock copolymer Pluranic P123 was used as structure directing agent, water as solvent, tetraethyl orthosilicate (TEOS) as silica source, and aluminum sulfate as aluminum source.

The mesoporous materials synthesized were characterized with BET, $NH_3$-TPD, TEM, $^{27}$Al and $^{28}$Si NMR, and ICP.

## 3. Results and Discussion

The effect of acid concentration and initial gel Si/Al ratio on surface areas, pore volumes, and pore sizes of the finial products is summarized in the table below. The results illustrated that the pore volume is decreased and the surface area increased with increased acid concentration. The surface area and pore volume are almost unchanged when the Si/Al molar ratio of the initial gel changed from 5 to 50.

| HCl concentration, M | Acid free | | | 0.5 | | |
|---|---|---|---|---|---|---|
| Initial Si/Al gel molar ratio | 5 | 20 | 50 | 5 | 20 | 50 |
| BET surface area, $m^2/g$ | 575 | 600 | 544 | 879 | 887 | 920 |
| Pore volume, ml/g | 1.2 | 1.23 | 1.24 | 1.04 | 0.97 | 0.98 |
| Average pore size, nm | 8.38 | 8.18 | 8.78 | 5.34 | 5.03 | 6.37 |

The pore size and wall thickness are shown in the Figure below. The result is well in agreement with BET results. The wall thickness dramatically decreased with the HCl concentration, and then level off when HCl concentration was higher than 0.5M.

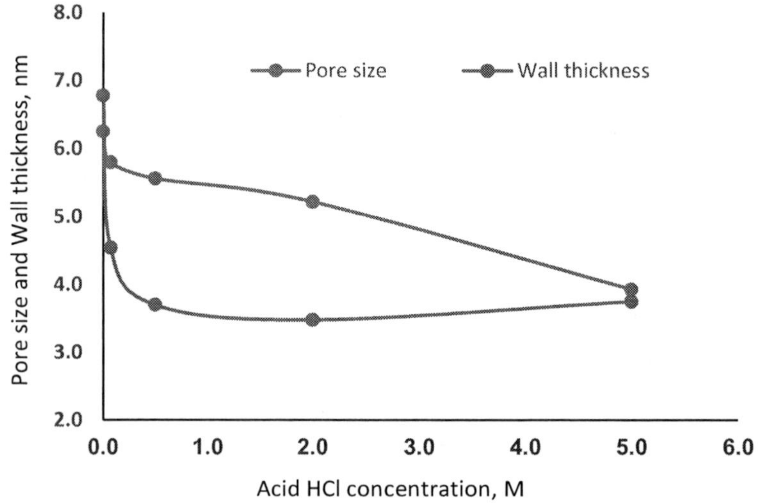


$^{27}Al$ NMR results clearly indicated that a considerable amount of tetrahedral Al existed when the SBA-15 was synthesized at low acid or acid-free environment.

The results of the effect of HCl concentration and initial Si/Al ratio on final product Si/Al molar ratio showed that the acidity (from 0 to 0.5M HCl) had no significant effect on final product Si/Al ratio.

## 4. Conclusions

The acid free and weak acid environment favored the enhancement of acidity, and the formation of larger mesoporous (higher pore volume, pore size, and surface areas).

A considerable amount of Al was in tetrahedral Al state, which help the further conversion to the zeolite wall.

The effect of the original gel Si/Al ratio on textural properties and acidity is not significant.

The Si/Al molar ratios are all above 50 when the initial Si/Al ratios changed from 5 to 50.

Paper No. 7

# Kinetics of simultaneous HDS of DBT and 4-MDBT/4,6-DMDBT over CoMoP/$\gamma$-$Al_2O_3$ catalysts

***Syed Ahmed Ali[1], Mohammad M. Hossain[2]***
*[1]Center of Research Excellence in Petroleum Refining and Petrochemicals,*
*[2]Department of Chemical Engineering, KFUPM, Dhahran 31261, Saudi Arabia*

A series of CoMo/$\gamma$-$Al_2O_3$ catalysts were prepared with an addition of 0.0, 0.5, 1.0 or 1.5% phosphorus pentaoxide to investigate the influence of phosphorus addition on the simultaneous hydrodesulfurization (HDS) of refractory sulfur compounds. Two sets of HDS experiments were carried out using sulfur bearing model compounds: (i) DBT (dibenzothiophene) with 4-MDBT (4-methyl dibenzothiophene) and (ii) DBT (dibenzothiophene) with 4,6-DMDBT (4,6-dimethyl dibenzothiophene). The Langmuir-Hinshelwood mechanism based kinetics model was developed from the experimental conversion and product distribution data for simultaneous HDS of the two sulfur compounds via direct desulfurization (DDS) and hydrogenation (HYD) routes. The kinetic model fits the experimental data quite adequately for all the species. The HDS rate for DBT was found to be approximately 2 and 7 times higher than that estimated for 4-MDBT and 4,6-DMDBT, respectively. Enhancement of HDS rate by P addition was observed up to 0.1 g/g (1.0 wt%) $P_2O_5$ while higher amount of P is not advantageous. A marginal preference towards HYD pathway was observed for the HDS of 4,6-DMDBT while DDS route was predominant during the HDS of DBT and 4-MDBT. The trends in values of the rate constants estimated by the Langmuir-Hinshelwood kinetic model compared well with pseudo-first rate constants.

Paper No. 8

# Development of a Super-critical Water Cracking Process for Pipeline Transportation of Extra Heavy Crude Oil

***Takayoshi Fujimoto, Hisato Aoyama, Tomoki Kayukawa***
*JGC, Japan*

One of the big issues for the development of extra heavy crude oil, such as Canadian oil sand bitumen, is the difficulty in transportation from the wellhead to the refinery because of its high density and viscosity. Conventionally, a dilution method which dilutes extra heavy oil with a diluent or a full-upgrading method, typically a combination of thermal cracking and catalytic processing, is applied to improve the density and viscosity to a level that will permit transport through pipelines. However, these conventional approaches cannot be applied for all cases for reasons related to the availability of a diluent source or from the economic point of view.

Partial upgrading is a new approach to produce pipeline transportable crude through the use of a simpler process configuration than the full-upgrading method.

JGC's Supercritical Water Cracking (SCWC) process is a partial upgrading process. A 5 bpd pilot facility has been successfully operated with Canadian oil sand bitumen.

In this study, an outline of the SCWC process is introduced together with information regarding its performance and a description of the characteristics of the products. The results of an economic evaluation are also shown to compare the SCWC process with the conventional methods.

# Development of a Supercritical Water Cracking Process as a Transportation Method for Extra Heavy Crude Oil

**Takayoshi Fujimoto[*1], Hisato Aoyama[*1], Tomoki Kayukawa[*1]**
*1: Technology Innovation Center, Technology Innovation Division, Infrastructure Division, JGC Corporation, 2-3-1, Minato Mirai, Nishi-ku, Yokohama 220-6001, JAPAN

## 1. Abstract

One of the big issues for the development of extra heavy crude oil, such as Canadian oil sand bitumen, is the difficulty in transportation from the wellhead to the refinery because of its high density and viscosity. Conventionally, a dilution method which dilutes extra heavy oil with a diluent or a full-upgrading method, typically a combination of thermal cracking and catalytic processing, is applied to improve the density and viscosity to a level that will permit transport through pipelines. However, these conventional approaches cannot be applied for all cases for reasons related to the availability of a diluent source or from the economic point of view.

Partial upgrading is a new approach to produce pipeline transportable crude through the use of a simpler process configuration than the full-upgrading method.

JGC's Supercritical Water Cracking (SCWC) process is a partial upgrading process. A 5 bpd pilot facility has been successfully operated with Canadian oil sand bitumen.

In this study, an outline of the SCWC process is introduced together with information regarding its performance and a description of the characteristics of the products. The results of an economic evaluation are also shown to compare the SCWC process with the conventional methods.

## 2. Introduction

Extra heavy oil is an important natural resource to help respond to the growth of oil demand. However, one big issue for the development of extra heavy crude oil is its transportation from the well head to the refinery or shipping facility, because of its high density and high viscosity. Currently three methods which are dilution technology, full upgrading technology, and partial upgrading technology are used to enable its transportation (**Figure 1**). Conventionally, dilution technology is applied to reduce density and viscosity. But the OPEX (operating expenditure) involved is high because of the cost of the diluent and the transportation. In addition, full upgrading technology to produce high quality SCO (synthetic crude oil) is also commercialized. However, the CAPEX (capital expenditure) and OPEX are high because the configuration is complex and hydrogen and a catalyst for hydrogenation are required for processing. Partial upgrading technology is a new approach to produce SCO from heavy oil. The difference from full upgrading technology is that the partially upgraded product is sour SCO which contains some amount of impurities, e.g. sulfur and nitrogen, but satisfies pipeline specifications in terms of density and viscosity. A Canadian provincial organization, Alberta Innovates, is focusing on the development of partial upgrading technologies and is targeting to treat up to 20 percent of its in-situ production by partial upgrading in 2030[1)]. Partial upgrading technology is expected to reduce or replace the diluent requirement for heavy oil transportation with lower OPEX and lower environmental risk.

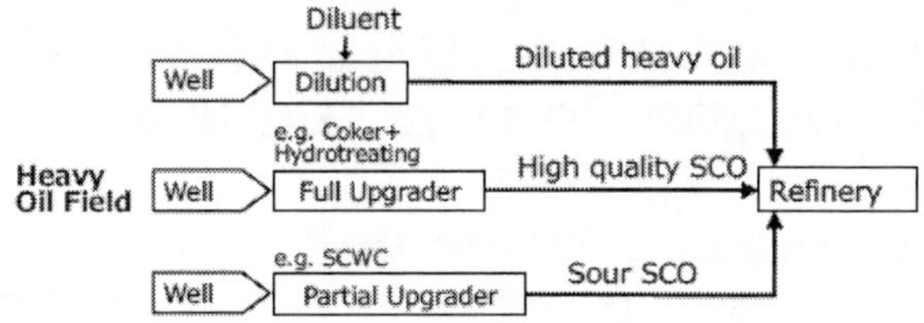


Figure 1 Methods to Transport Heavy Oil from Wellhead to Refinery

JOGMEC (Japan Oil, Gas and Metals National Corporation) and JGC have been developing the SCWC (Supercritical Water Cracking) process as a partial upgrading technology since 2006 to make a contribution to the development of unconventional oil fields. The SCWC process is categorized as thermal cracking technology. Some fundamental researches play important roles to establish the thermal cracking process with supercritical water[2),3)]. Whatever thermal technology is used for heavy oil conversion, the cracked products contain some amount of olefins. Thus the products may be unstable and not acceptable to the refineries.

In this paper, the characteristics and performance of the SCWC process are described. Then, the technical evaluation of product stability for partially upgraded heavy product of the SCWC process is addressed. Compatibility with other crudes and long-term storage stability were examined to describe product stability during storage and transportation.

The economic evaluation is also shown. Supply costs of Canadian oil sand bitumen from upstream wellhead, midstream pipeline and downstream refinery were compared for three cases, i.e. dilution, full upgrading and partial upgrading. The supply costs and product values for these three cases were compared, and the sales margins were also calculated.

## 3. Outline of the SCWC process

### 3.1 Technical Features

**Figure 2** shows a schematic diagram of the SCWC reaction.

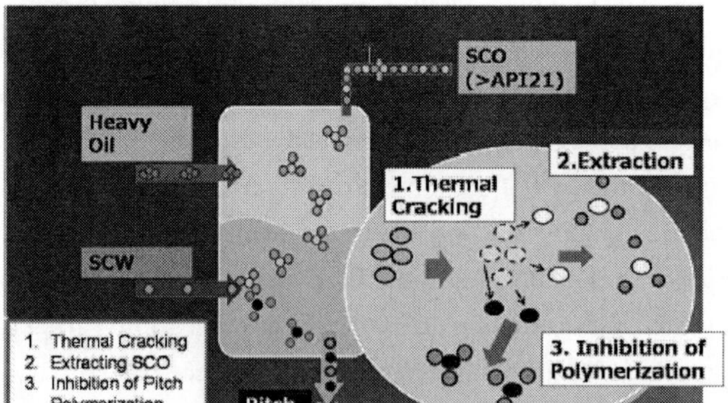


Figure 2 Schematic Diagram of SCWC Reaction

Thermal cracking occurs due to heat input from the supercritical water in the reactor and SCO is extracted immediately from the reactor by the up-flowing supercritical water. Therefore, excess cracking reaction of the SCO is prevented and this results in lower cracked gas yield. The supercritical water also has the characteristic of preventing polymerization of heavy oil molecules[3)]. Based on these technical features, a high yield of SCO and low yields of pitch and gas can be achieved.

### 3.2 Process Description

A simplified flow diagram of the SCWC process is shown in **Figure 3**. The SCWC process is applied for upgrading to produce SCO from a wide range of heavy oil or high pour-point oil sources such as Bitumen, AR (atmospheric residue), VR (vacuum residue) and pitch from SDA (solvent deasphalting).

The feedstock is pressurized by a feed pump and passed through a preheater. At the same time, water is also pressurized by a water pump and preheated above the critical temperature of water by heaters. Preheated feedstock and supercritical water are introduced from the top and the bottom of the reactor, respectively. The ratio of feedstock and water is typically 1:1 by weight. The typical reaction temperature and pressure in the reactor are 395–425 °C and 22–25 MPaG, respectively.

In the reactor, thermal cracking occurs by heat input from the supercritical water. And SCO is extracted immediately from the reactor by the up-flowing supercritical water. A mixture of SCO and supercritical water is recovered from the top of the reactor and pitch is recovered from the bottom.

A high pressure separator and a low pressure separator are installed for the separation of SCO, cracked gas and sour water from the reactor effluent. The cracked gas is used for fuel after the removal of acid gas. Sour water is fed through the sour water stripper and then introduced to the waste water treatment unit. Treated water is recycled as feed water.

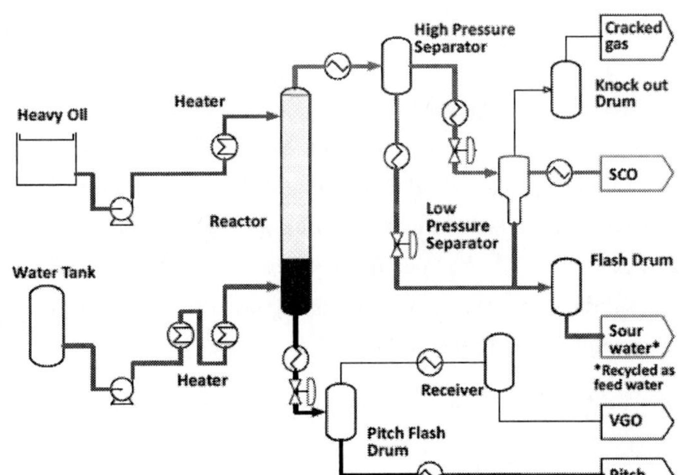


Figure 3 Flow Diagram of SCWC Process

## 4. Experimental

### 4.1 Pilot Plant Operation

Experimental runs with the SCWC pilot plant were conducted to confirm the performance with Canadian oil sand bitumen as the feedstock. **Table 1** shows the properties of the feedstock, and **Table 2** shows the typical operating condition.

Table 1 Properties of Feedstock

| Properties | Unit | | Method |
|---|---|---|---|
| API gravity | [°] | 8.3 | ASTM D70 |
| Kinetic Viscosity at 10degC | [cSt] | >600,000 | ASTM D5018 |
| Distillation | | | |
| < 360degC | [wt%] | 16 | ASTM D7169 |
| 360-540degC | [wt%] | 32 | ASTM D7169 |
| > 540degC | [wt%] | 52 | ASTM D7169 |
| Sulfur | [wt%] | 5.2 | ASTM D4294 |
| MCR | [wt%] | 14.8 | ASTM D4530 |

Table 2 Typical Operating Condition

| Conditions | Unit | |
|---|---|---|
| Temperature | [degC] | 425 |
| Feedstock Flow Rate | [BPSD] | 4.7 |
| Water Flow Rate | [BPSD] | 5.0 |

### 4.2 Product Stability Test

#### 4.2.1 Compatibility

Compatibility is a key characteristic to discuss the asphaltene stability of a crude oil when blending more than two kinds of crude oils. If compatibility is good, the asphaltene in the crude oil stays soluble, but if compatibility is poor, the asphaltene in the crude oil could precipitate and create solid sludge. The partially upgraded product is usually expected to blend with other crude oil at a blending facility and stored at a product pool before transportation through a pipeline. Therefore, the compatibility between the SCO produced by the SCWC process (hereinafter called SCWC SCO) and other crude oils should be evaluated. In this study, Western Canadian Select (WCS, 21.6°API), which is a heavy crude oil commercially available in Canada, was selected as a blending oil.

Dr. Irwin A. Wiehe's compatibility model and his testing method were applied for compatibility evaluation[4)]. In this method, two key parameters, insolubility number ($I_N$) and solubility blending number ($S_{BN}$) of each blending oil, were measured by two kinds of tests, heptane dilution test and toluene equivalent test.

The heptane dilution test is a test to determine the maximum volume of n-heptane ($V_H$) that can be blended with 5 ml of a sample oil without precipitation of asphaltene. The toluene equivalent test is a test to blend 2 grams of sample oil and 10 ml of test liquid. By varying the volume% of toluene mixed with n-heptane in the test liquid, the minimum volume % of toluene (TE) required to keep the asphaltene soluble is determined. The key parameters, $I_N$ and $S_{BN}$ of sample oils are determined by **Eq. 1** and **Eq. 2**. A solubility blending number of blend oil ($S_{BNmix}$) is calculated by the volumetric average of the components, defined by **Eq.3**.

$$I_N = \frac{TE}{\left(1 - \frac{V_H}{25 \times d}\right)} \quad \text{... (Eq.1)}$$

$$S_{BN} = I_N \times \left(1 + \frac{V_H}{5}\right) \quad \text{... (Eq.2)}$$

Where;
$I_N$ is the insolubility number
$S_{BN}$ is the solubility blending number
$V_H$ [ml] is a number determined by the heptane dilution test
TE [vol%] is a number determined by the toluene equivalent test
d [g/ml] is the density of the sample oil

$$S_{BNmix} = \frac{V_1 S_{BN1} + V_2 S_{BN2} + \dots}{V_1 + V_2 + \dots} \quad \dots \text{(Eq.3)}$$

When the $S_{BNmix}$ is higher than the insolubility number of all the blended oil, the blended oil is compatible, meaning that it is stable without asphaltene precipitation. Thus, the compatibility criterion is defined by **Eq.4** with the maximum value of $I_N$ ($I_{Nmax}$)

$$\text{Compatibility criterion: } S_{BNmix} > I_{Nmax} \quad \dots \text{(Eq.4)}$$

### 4.2.2 Storage Stability

Storage stability is a characteristic to determine that no sludge is created in a tank when the oil is stored for a long period. Various standard methods are defined by the American Society for Testing and Materials (ASTM). However, no standard method is defined to evaluate long term stability for SCO because SCO is an interim product, not a final product. In this study, two kinds of ASTM method, ASTM D4625 and D4870, were applied to evaluate the long term storage stability of SCWC SCO.

## 5. Results and Discussions

### 5.1 Results of Pilot Plant Operation

**Table 3** shows the VR (distillation range > 540°C) conversion and product yield. SCO yield was above 75 vol% if the Pitch vacuum flasher was installed downstream of the Pitch Flash Drum in Figure 3. In spite of gas generation (1.1 wt%), the total liquid yield was 99.5 vol%. Such a high liquid yield is one of the characteristics of the SCWC process.

Long term operation was also conducted to confirm the reliability of the process. During long term operation, the properties of the products were stable and there was no trouble. Reliability of operation of the SCWC reactor was confirmed during this testing period.

Table 3 VR Conversion and Product Yields by Pilot Plant

| Items | Unit | |
|---|---|---|
| VR conversion (SCWC Unit) | [%] | 38.4 |
| Product Yield (SCWC Unit) | | |
| Off Gas | [wt%] | 1.1 |
| Synthetic Crude Oil | [vol%] | 61.4 |
| Pitch | [vol%] | 38.1 |
| Total Liquid | [vol%] | 99.5 |
| Product Yield (SCWC Unit + Pitch Vacuum Flasher) | | |
| Synthetic Crude Oil | [vol%] | 75.2 |
| Pitch | [vol%] | 24.3 |

**Table 4** shows the properties of SCWC SCO at the typical operating condition. The target of SCO qualities is to satisfy pipeline specifications. For example, Canadian pipeline specifications require the density to be lower than 940 kg/m3 (API° 19), kinetic viscosity to be lower than 350 cSt at 10°C, and the olefin content to be lower than 1.0 wt% (n-decene equivalent)[5]. The API gravity and the kinetic viscosity of SCWC SCO satisfy these specifications. The sulfur content did not change drastically by the SCWC process. The MCR (micro carbon residue) content was reduced to less than 1 wt%. The olefin content of SCWC SCO is greater than the amount stipulated in Canadian pipeline specifications. To achieve the specification of olefins, further treatment, such as hydrotreating, is required. The diene number of SCWC SCO is 3.24, which is not low enough, but lower than that of the coker product, which is 4~9. Therefore, there is a low possibility of operational problems at the downstream processing, such as storage, transportation and refinery processing. The technical evaluation of SCWC product quality is discussed in the next section.

Table 4 SCWC SCO Properties

| Properties | Unit | | Target | Method |
|---|---|---|---|---|
| API gravity | [°] | 24.3 | > 19 | ASTM D4052 |
| Kinetic Viscosity at 10degC | [cSt] | 23 | < 350 | ASTM D7042 |
| Olefin | [wt%] | 4.4 | < 1.0 | 1H-NMR |
| Diene Value | [g-$I_2$/100g] | 3.24 | | UOP 326 |
| Distillation | | | | |
| < 360degC | [wt%] | 56 | | ASTM D2887 |
| 360-540degC | [wt%] | 40 | | ASTM D2887 |
| > 540degC | [wt%] | 4 | | ASTM D2887 |
| Sulfur | [wt%] | 3.2 | | ASTM D4294 |
| MCR | [wt%] | 0.5 | | ASTM D4530 |

**5.2 Product Stability Test**

**5.2.1 Compatibility**

**Table 5** shows the analytical results of $I_N$ and $S_{BN}$ for SCWC SCO and WCS. As SCWC SCO does not contain asphaltene, $I_N$ is zero. $S_{BN}$ for SCWC SCO was 59.1, which is greater than the typical value for the paraffinic solvent, and is close to the value for the aromatic solvent. With respect to WCS, $I_N$ was 33.1, and $S_{BN}$ was 93.3. $S_{BN}/I_N$ of WCS was 2.82, asphaltene in WCS is definitely stable. $S_{BNmix}$ for the blend oil of SCWC SCO and WCS is calculated at each blending ratio by Eq.3. $I_{Nmax}$ is 33.1 for the mixture of SCWC SCO and WCS. **Figure 4** shows the relationship between the blending ratio of SCWC SCO over WCS and $S_{BNmix}$, $I_{Nmax}$. $S_{BNmix}$ is greater than $I_{Nmax}$ at all blending ratios. It means that the blended oil is compatible by Eq.4.

Table 5 Compatibility Numbers of SCWC SCO and WCS

| | SCWC SCO | WCS |
|---|---|---|
| $I_N$ | 0 | 33.1 |
| $S_{BN}$, $S_{SO}$ | 59.1 | 93.3 |

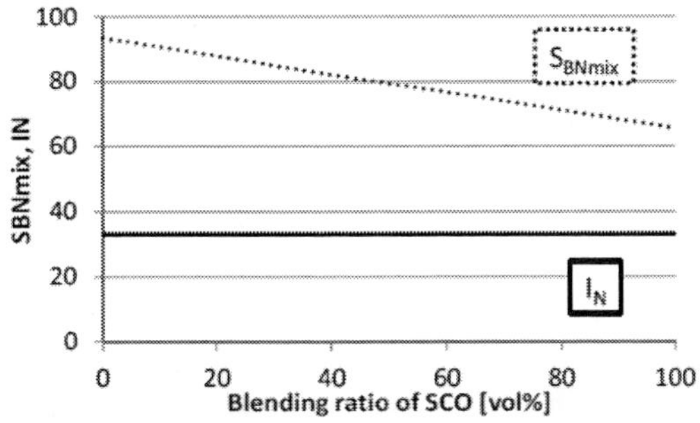


Figure 4 Relationship between Blending Ratio of SCWC SCO with WCS and $S_{BNmix}$, $I_{Nmax}$

### 5.2.2 Storage Stability

From the point of pipeline specifications, 0.5% of BS&W (Base Sediment and Water) is only a related criterion. From the operational point of view, 0.1% of sludge formation is considered as a criterion to prevent operational problems. **Figure 5** shows the historical trend of sludge content at the end of each period. The sludge content was increasing by the end of week 12, and it kept stable after week 12. The sludge of SCWC SCO was less than 0.03% at its maximum as shown in Figure 5, so SCWC SCO has sufficient storage stability.

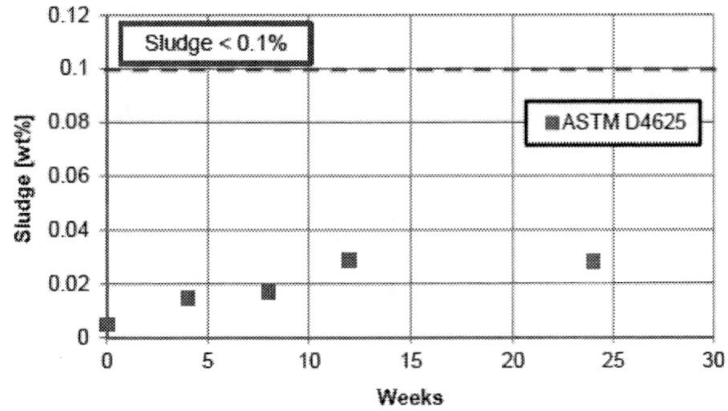


Figure 5 Long Term Storage Stability of SCWC SCO (ASTM D4625)

**Figure 6** shows the results of the determination of the total sediment for each sample. The total sediment of SCWC SCO and the mixture of SCWC SCO and WCS were 0.01wt%. This number was equivalent to the results for the reference oils, the total sediment of Dilbit was 0.02wt% and that of Synbit was 0.01wt%. As described above, 0.1% of sludge formation is considered as a criterion which does not cause operational problems. The SCWC SCO and the blended oil of SCWC SCO and pitch or WCS are considered as stable in terms of storage stability.

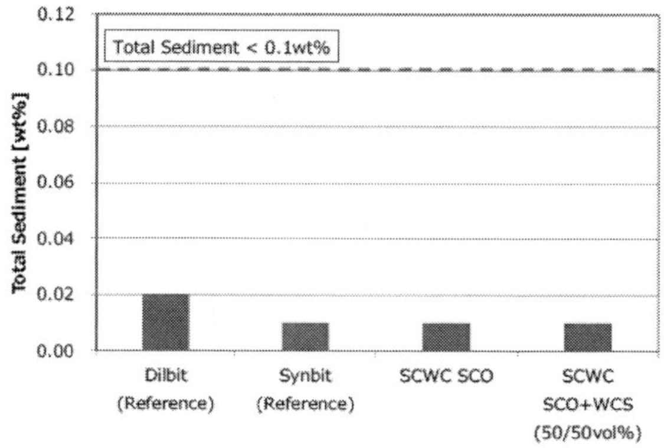


Figure 6 Total Sediment of SCWC SCO (ASTM D4870)

### 5.3 Economic Evaluation

As described in the introduction to this paper, three kinds of cases were studied for bitumen transportation. In this chapter, the economics of partial upgrading by the SCWC process were evaluated in comparison to the dilution and the full upgrading. The capacity of bitumen production was 30,000 BPD in each case.

#### 5.3.1 Block Flow

**Figure** 7 shows the block flow diagrams of the bitumen transportation, the dilution case, Full Upgrading Plant, and the SCWC plant.

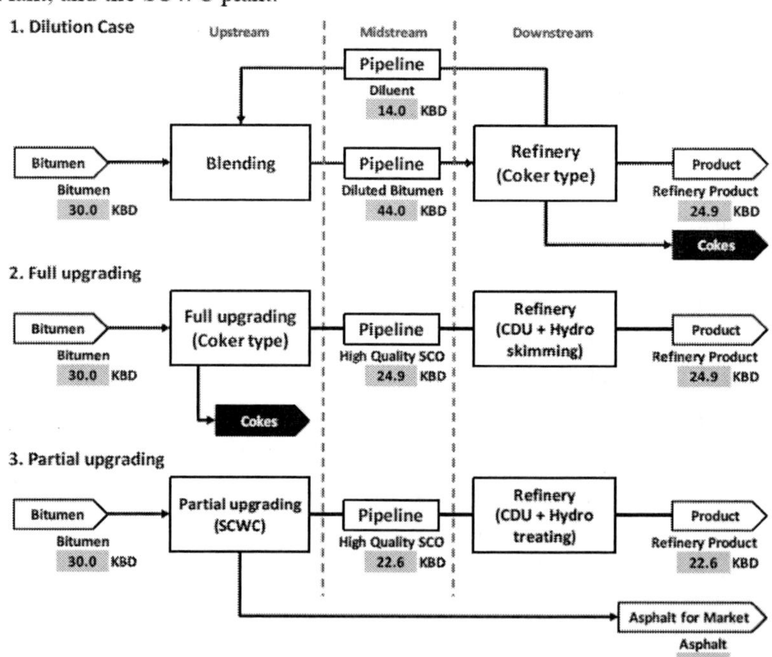


Figure 7 Block Flow Diagram for the Bitumen Transportation

(1) Conventional Approaches -Dilution Case -

In upstream, the bitumen is produced and blended with diluent to reduce the density and viscosity of the bitumen. Diluted bitumen is transported by pipeline, and then diluent is recovered at the refinery and bitumen is processed by a refinery with coker. Diluent recovered at the refinery is transported to the blending facility at the wellhead for reuse. In this case, a complex facility is not required for the upstream facilities. However, the pipeline needs to have a larger capacity (44,000 BPD) to cope with the combined volume of bitumen (30,000 BPD) and diluent (14,000 BPD), and an additional pipeline for diluent recycling is also required. The diluted bitumen can be processed only by refineries which have a coker or other bottom upgrading process.

(2) Conventional Approaches -Full upgrading-

The bitumen is processed by a full upgrading facility and then high quality SCO, e.g. low sulfur and no residue, is produced. The SCO is transported through a pipeline, and refinery products can be obtained by a hydro skimming type refinery. The required pipeline capacity is 24,900 BPD. However the upgrader at upstream is so complex that the operating cost is high due to hydrogen consumption and catalyst for hydrogenation as well as high investment cost. The cokes are disposed of at the upstream wellhead because they have no commercial value there, and it is not preferable from the environmental point of view.

(3) New Approach -Partial Upgrading-

SCWC is a partial upgrading technology to produce SCO near the wellhead by means of a simple facility. The partially upgraded product is not the same high quality as the fully upgraded product in terms of sulfur content. However, the product satisfies pipeline specifications in terms of density and viscosity, so it can be transported without dilution, then the pipeline for diluent is not required. The by-product of pitch can be considered to have value as asphalt.

**5.3.2 Assumptions**

The assumptions in this paper are summarized below.

<Feedstock cost>

・WTI price : $60/bbl ・Diluent : $65/bbl

<Upstream>

・CAPEX of SAGD : $6.3/bbl-Bitumen ・OPEX of SAGD : $8.8/bbl-Bitumen

<Midstream>

・Pipeline cost : $10/bbl

<Utility in upstream and downstream>

・Power : ¢7.5/kWh ・Natural Gas Price : $4/MMBTU

<Product price>

・SCO : $60/bbl ・Cokes : $30/ton ・Road asphalt : $40/bbl

### 5.3.2 Results of Economic Evaluation

**Figure 8** shows the results of an economic evaluation for each case. From the left, the dilution case, full upgrading case and partial upgrading case with the SCWC process were shown. For each case, the left side bars show the supply cost per barrel of bitumen, including SAGD CAPEX, SAGD OPEX, Natural gas for SAGD, upgrader cost, pipeline cost, pipeline cost (return) and refining cost at the downstream refinery. The right side bars show product value produced from a barrel of bitumen. The margin can be calculated from the difference between the supply cost and product value in each case.

In comparing the supply cost, the dilution case shows the highest number due to higher pipeline cost. The supply cost for the SCWC case is the lowest number due to the lower upgrader cost and the lower pipeline cost.

In comparing the product value, the diluent case and the full upgrading case are the same because both cases apply coker in downstream or upstream. The SCWC case shows the highest value of product sales, because the pitch was valuable by-product as asphalt.

The margin of the SCWC case was \$23.2/bbl-bitumen while the margin was \$16.8/bbl-bitumen for the full upgrading case and \$8.6/bbl-bitumen for the dilution case. Supposing that there is a good market for road asphalt in North America, hence the SCWC process would be the preferred solution compared to conventional approaches.

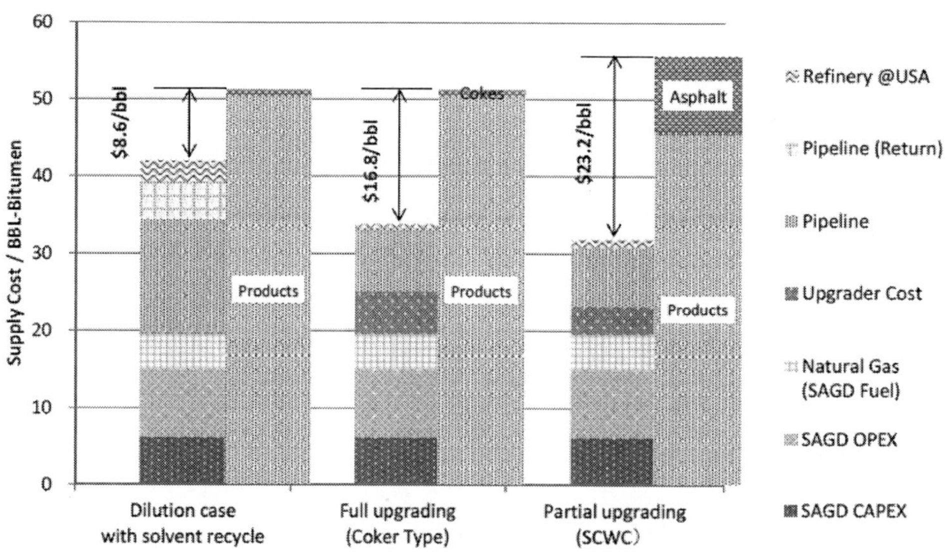


Figure 8 Supply Cost and Product Values for Each Application

### 6. Conclusion

The SCWC process is a partial upgrading process which produces pipeline transportable oil from heavy oil. The SCWC process has advantages in lower operating cost by eliminating diluent costs for pipeline transportation and lower capital cost due to its simple process configuration, and minimal waste product.

The SCWC process was demonstrated with a 5 BPD scale pilot plant. The reliability of the process was demonstrated at the pilot plant, and the stability of the product was analysed and examined with laboratory scale units in terms of compatibility and storage stability.

The economics of the SCWC process were compared with the conventional dilution case and full upgrading case. The margin shown in the SCWC case was identified to be higher than for the other cases.

JGC continues to work on the SCWC process for its application to various kinds of heavy oil. The SCWC process as a partial upgrading technology or field upgrading technology can provide a cost effective solution for heavy oil transportation.

**7. Acknowledgement**

This research and development program has been co-developed with JOGMEC (Japan Oil, Gas and Metals National Corporation). The pilot plant operations of the SCWC and product analysis were conducted through cooperation with Natural Resources Canada's CanmetENERGY lab in Devon, Alberta, Canada. The study on product stability was conducted with the services of Omnicon Consultant Inc.

**8. References**

1) Alberta Innovates, "Enhancing Alberta's competitiveness: National Partial Upgrading Program" Website of Alberta Innovates. Available at http://www.ai-ees.ca/partial-upgrading-program/

2) JPEC, Japan Petroleum Energy Center, "R&D for Creation of Technological Seeds of Innovative Refining" Website of JPEC. Available at http://www.pecj.or.jp/english/technology/technology05.html, or research papers are available at http://www.pecj.or.jp/japanese/report/tech_thema/rep01_s.html#s4

3) Morimoto, M., Sugimoto, Y., Saotome, Y., Sato, S., Takanohashi, T., 2010. "Effect of supercritical water on upgrading reaction of oil sand bitumen". the Journal of Supercritical Fluids, 55, (1), 223-231: Elsevier B.V.

4) Irwin A. Wiehe, 2008, Process Chemistry of Petroleum Macromolecules, 205-217, Boca Raton, FL: CRC Press Taylor & Francis Group

5) Enbridge Inc. "Enbridge Service Levels (June 2015)". Website of Enbridge Inc. Available at http://www.enbridge.com/~/media/Enb/Documents/Shippers/Enbridge_Mainline_Service_Levels.pdf?la=en

Paper No. 9

# Hydrodynamics and solids mixing in fluidized beds with inclined-hole distributors

***Alhussain Bakhurji, Xiaotao Bi, John R. Grace***
*Department of Chemical and Biological Engineering,*
*University of British Columbia, Vancouver, Canada V6T1Z3*

Fluidization is a process in which solid particles are suspended in a fluid-like form subject to several forces acting on them. These forces include gravity, buoyancy, and drag. Particles are grouped according to their mean particle size and density in a classification proposed by Geldart (1973) who classified these particles into four different groups (A, B, C and D) for particle densities of typical particles under room temperatures and pressure. Group A (Aeratable) has an average particle size of ~ 30-100 microns. Group B (Bubbling) has an average particle size of ~ 100-1000 microns; whereas, Group C (Cohesive) has an average particle size of less than ~ 30 microns and Group D (Spoutable) has an average particle size of more than ~ 1000 microns.

Fluidized bed reactors can have many configurations, depending on the process being applied. There is the circulating fluidized bed reactor in which solids mixing is improved by circulating the solids back to the bottom of the reactor after passing through a riser. Such processes usually operate at high superficial gas velocity in the fast fluidization flow regime. It is also considered to be a type of reactor with good fluid-to-particle heat and mass transfer.

In the configuration of an annular fluidized bed, particles are fluidized between concentric outer and inner columns. This type of reactor can combine the benefits of conventional fluidized beds (long residence time) and the circulating fluidized bed (good heat and mass transfer) (Collin et al., 2009).

Another interesting type of fluidized beds is the swirling fluidized bed, where the fluidizing gas is introduced at an inclined angle at the distributor so that there is a tangential component, and the gas swirls inside the fluidization zone. The gas flow can then be divided into two components, a vertical component that is responsible for supporting the particles weight-minus-buoyancy, and a horizontal component that induces swirling motion (Sreenivasan et al., 2002). Such reactors typically have a central cone or cylinder that limits the utilization of volume inside the reactor. It is like the annular fluidized bed except for the distributor design in which there can be inclined holes or an annular spiral distributor or any other type with an angle other than 90° to the horizontal axis. Some applications of swirling fluidized beds include drying and combustion processes using Geldart type D particles (Mohideen et al., 2012).

In a swirling fluidized bed, the bed height is an important factor in determining the bed behavior. A shallow bed usually swirls as a single mass; whereas, a deep bed behaves differently. As the bed height increases, two layers of fluidization occur; a bottom layer that is swirling and a top layer that is bubbling (Kaewklum et al., 2009).

A swirling fluidized bed has some advantages over conventional fluidized beds in which there is no swirl. In a swirling bed, due to the inclined angle of gas injection, the process can operate at higher superficial velocity with less elutriation compared to a conventional bed. This is due to the gas injection angle that reduces the vertical component of the gas flow compared to a conventional fluidized bed in which all entering gas flows vertically. The centrifugal force also pushes particles to the outer wall of the column, where the upward gas velocity is lower (Sreenivasan et al., 2002). A disadvantage of the swirling bed is that with increasing air velocity, the swirling affects the particle movement where the particles migrate from the center to the outer wall. This may create an empty space adjacent to the inner wall of the column in which gas bypassing can occur.

An application that could benefit from the swirling fluidized bed is the decoupled adsorption-reduction selective catalytic reduction (SCR) process for Nitrogen Oxides (NOx). In this process, the flue gas enters the reactor in one region, while the reductant gas enters in an adjacent region in such a way that catalyst particles flow tangentially from an area of adsorption to an area of reduction. An issue found in this application (Yang, 2009) is that the reactor performance was limited by the catalyst adsorption capacity since the overall NOx conversion increased with decreasing gas velocity in the annulus. Therefore, to improve the catalyst performance, a low gas velocity in the region for adsorption (increasing the catalyst residence time) could be used (Yang, 2009).

Most previous research on swirling fluidized bed reactors did not use Group A/B particles that are common in catalytic reaction applications. This research on swirling fluidized beds with Group A/B particles focused on investigating the impact of introducing gas flow at an angle at the distributor on the overall residence time and mixing of the particles in the reactor. Specifically, how would changing the inclination angle of orifices in the distributor affect the pressure drop across the distributor and over the bed, the minimum fluidization velocity, the tangential and vertical mixing of particles and bed expansion?

Three distributors were designed and fabricated, all with the same specifications with the exception of the orifice inclination angle. The angles were vertical (90°), 45° and 30°. Two bed materials were used, glass beads and phosphorescent particles, belonging to Groups A and B in the Geldart classification. Two studies were conducted, a hydrodynamic study and a solids mixing study. In the hydrodynamic study, the effects of orifice inclination angle and static bed height were investigated. In the solids mixing study, the axial solids mixing was investigated for the 90°-hole distributor using phosphorescent tracer particles and by estimating the bed turnover time and dense-phase downward velocity. In addition, tangential solids mixing was studied for all three distributors using the phosphorescent tracer particles to measure the residence time distribution to estimate the tangential dispersion coefficient and the tangential particle velocity using the dispersion model. It was found that for a given superficial velocity, the distributor pressure drop was lower for the 45°-hole and 30°-hole distributors than the 90°-hole distributor.

The magnitude was around 10% less than for the 90°-hole distributor at the same superficial gas velocity. Bed pressure drop was lowest for the 90°-hole distributor; whereas, there was no clear difference between the 45°-hole and 30°-hole distributors in terms of bed pressure drop. Bed pressure drop fluctuations were highest for the 90°-hole distributor, indicating larger bubbles. In a shallow bed, bed expansion was highest for the

90°-hole distributor followed by the 45°-hole and then the 30°-hole distributor. However, in a deep bed, bed expansion was virtually the same for all three distributors. The minimum fluidization velocity was highest for the 30°-hole distributor followed by the 45°-hole and then the 90°-hole distributor. Static bed height had an impact on the minimum fluidization velocity for the inclined-holes distributors. With regards to solids mixing, the turnover time for the 90°-hole distributor could be estimated based on the bubbling bed model equations. It was concluded that probes and ports in this study interfered with the solids mixing by reducing the dense phase downward velocity by around 40%. The tangential particle velocity increased with increasing radius in the annulus. The velocity was higher near the outer wall of the column and lower near the inner wall and it was highest for the 30°-hole distributor. Finally, both inclined-holes distributors (30° and 45° holes) resulted in greater tangential mixing than the 90°-hole distributor.

**References:**

Collin, A., K. -E Wirth, and M. Stroeder. 2009. Characterization of an annular fluidized bed. Powder Technology 190 (1): 31-5.

Geldart, D. 1973. Types of gas fluidization. Powder Technology 7 (5): 285-92.

Kaewklum, R., V. I. Kuprianov, and P. L. Douglas. 2009. Hydrodynamics of air–sand flow in a conical swirling fluidized bed: A comparative study between tangential and axial air entries. Energy Conversion and Management 50 (12): 2999-3006.

Mohideen, M.F., B. Sreenivasan, S.A. Sulaiman, and V.R. Raghavan. 2012. Heat transfer in a swirling fluidized bed with geldart type-D particles. Korean Journal of Chemical Engineering 29 (7): 862-7.

Sreenivasan, B., and V.R. Raghavan. 2002. Hydrodynamics of a swirling fluidised bed. Chemical Engineering & Processing: Process Intensification 41 (2): 99-106.

Yang, T., and X. Bi. 2009. Novel fluidized bed reactor for integrated NOx adsorption-reduction with hydrocarbons. Environmental Science & Technology 43 (13): 5049-53.

Paper No. 10

# Zeolite-based Catalyst Performance for Catalytic Reforming

***Mohammad Al Rebh, Ali Al-Shareef***
*Saudi Aramco R&DC*

Catalytic reforming is one of the most important refining processes that boosts the octane for gasoline, produces aromatics including BTX, and supplies hydrogen to hydrotreating processes. A number of reactions take place during reforming including dehydrogenation, isomerization, dehydrocyclization, and cracking that require both metallic and acidic sites on a catalyst. The typical commercial catalysts are alumina-based with platinum as the active metal with acidic sites maintained by continuous injection of chlorinated chemical with the feed. The objective of this work is to test a zeolite-based catalyst for catalytic reforming and compare its results with a typical commercial catalyst.

The zeolite-based catalyst was developed in-house for different application with an active metal that can theoretically perform as a catalytic reforming catalyst. The test was carried out in one of Saudi Aramco's Research and Development Center (R&DC) facility's pilot plants under various testing conditions to check for the catalyst stability and rate of deactivation, the effect of hydrogen to hydrocarbon ratio, pressure, and liquid space velocity. The range of conditions was chosen based on typical industrial reforming conditions. Results showed excellent reforming activity in terms of RON and aromatic content for the fresh catalyst as seen in **Figure 1**.

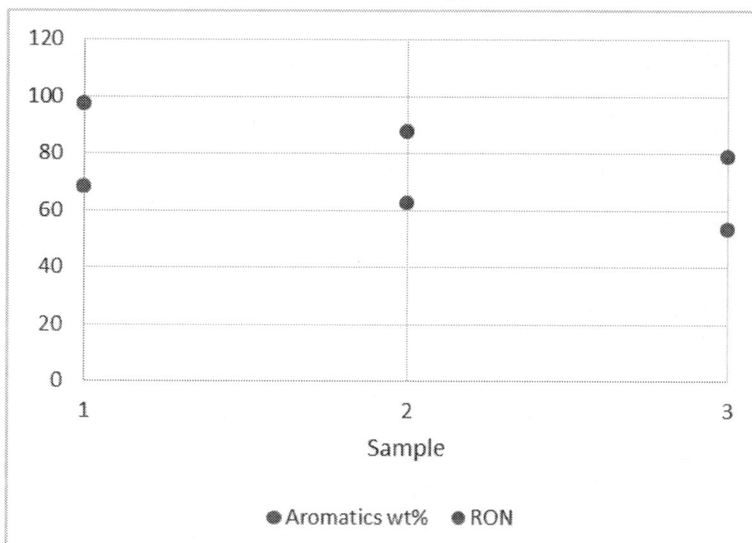


**Figure 1. Fresh zeolite-based catalyst performance at 510 °C**

Each condition was maintained for at least three days during which a daily mass balance was performed and the product collected was analyzed. For reference, an industrial facility operating a continuous catalytic reformer (CCR) at an average bed temperature (WABT) of

500°C produces reformate with a RON of 97. To compare the performance using the same pilot plant, the results were compared with a commercial catalyst tested in the same pilot plant and at similar conditions. The commercial catalyst test started at 490°C, 500°C, and then 510°C, with very little signs of deactivation; whereas, the zeolite-based was tested directly at 510°C. The two catalysts had similar initial activity at 510°C; but, the zeolite-based catalyst showed much quicker deactivation than the commercial catalyst, as seen in **Figure 2**.

Furthermore, the quick deactivation in the zeolite-based catalyst is likely due to a relatively higher degree of cracking and coking reactions than the commercial catalyst since zeolites typically induce a high rate of cracking in hydrocarbons at elevated temperatures. Due to this quick deactivation, full and accurate interpretation of the effect of change of parameters, such as pressure or liquid space velocity, was not indicative for the catalyst performance. Therefore, comparison will be limited at the initial condition when the catalyst was fresh.

**Table 1** shows different yields comparison between the first mass balances taken at 510°C for both catalysts, where both yielded 97 RON value. Although there are some errors in the gas products' yields for the commercial catalyst due to the 93% mass recovery, the feed composition and analysis are still comparable. In addition, the gas yields in **Table 1** support the conclusion that more cracking reactions are taking place on the zeolite-based catalyst.

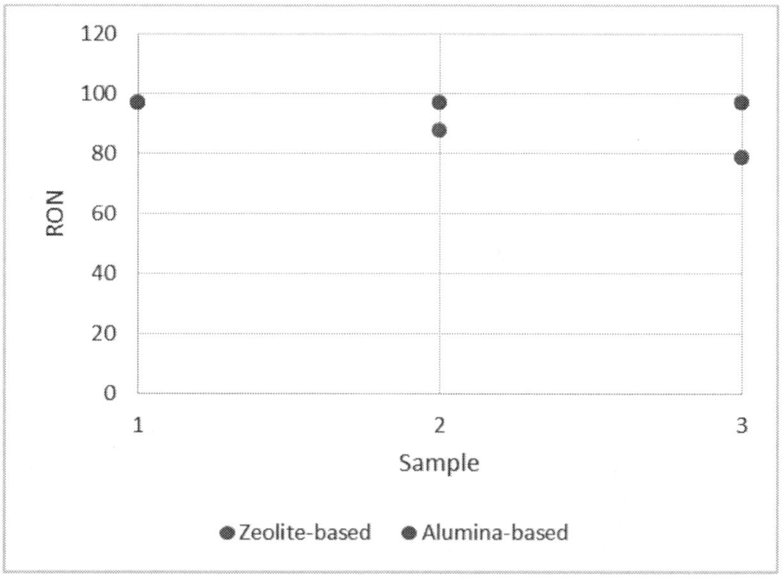


**Figure 2. Comparison in RON value for zeolite-based catalyst and Alumina-based catalyst at 510 °C**

The results indicate that there is a promise of having a zeolite-based reforming catalyst but requires further investigation to adjust its acidic activity and rate of deactivation. Even more, the fresh zeolite-based catalyst performance suggests the catalyst ability to perform as a reforming catalyst if it can be regenerated continuously, such as in a continuous catalyst recycle reforming process like CCR, where the catalyst is always regenerated.

**Table 1. Comparison in yield values for the first mass balance of the two catalysts at 510 °C**

| | Zeolite-based | Alumina-based |
|---|---|---|
| **Calculated RON** | 97.1 | 97.4 |
| **Aromatic content (wt%)** | 68.3 | 68.3 |
| **Olefins (wt%)** | 4.3 | 2.4 |
| **H2 Yield, wt%** | 0.55 | 0.65 |
| **Recovery wt% (Mass balance)** | 100% | 93% |
| **C5+ liquid yield, wt%** | 74.7 | 77.5 |
| **C3+C4 yield, wt%** | 18.6 | 9.2 |
| **C1+C2 yield, wt%** | 6.2 | 4.6 |

In conclusion, the zeolite-based catalyst could be beneficial in reforming heavier-cut naphtha and might not require chloride addition to sustain its activity due to its higher cracking capability. The initial finding of this experiment encourages subsequent investigation of zeolite-based reforming catalysis.

Paper No. 11 (Keynote)

# Petrochemicals – Technical and Economic Challenges for This Century

***Patrick C Whitchurch***
*Honeywell UOP, Des Plaines IL, USA*

By the middle of this century, world population is expected to be more than 9 Billion people. Worldwide energy demand will have increased by more than 35%, with gas growing to become over 25% of the world's energy consumption. There will be continued strain on fresh water resources as well as pressure on CO2 emissions. Along with these mega-trends, the rising middle classes of High Growth Regions will drive demand increases for Polypropylene, Polyethylene and PET to more than 150% of demand that is seen in 2017. Cost-advantaged feedstocks will continue to dominate the technologies used for these petrochemical building blocks.

Over this time, Chemical Engineers will be increasingly challenged to discover and develop technologies that provide the highest return on capital, the lowest consumption of energy per ton of product, and the most efficient use of available feeds, while meeting increasingly stringent environmental standards. This presentation gives an overview of these mega-trends and discusses some significant advances in petrochemical technologies that help meet these worldwide needs.

Paper No. 12

# Catalytic copolymerization of $CO_2$ and diols using $CeO_2$ and 2-cyanopyridine

***Keiichi Tomishige***
*Tohoku University, Japan*

Direct and catalytic polymerization of $CO_2$ and diols to polycarbonate is a promising technology as a simple and environmental-benign method. Generally, the synthesis of organic carbonates from alcohols and $CO_2$ needs selective catalysts and effective dehydrating agents because of the serious equilibrium limitation. One of the conventional synthesis methods of organic carbonates is the reaction of alcohols with phosgene. However, the problems of this phosgene method are the usage of poisonous phosgene and large amount of salt byproduct for the neutralization. Therefore the impact of the present work is related to the substitute of phosgene with $CO_2$. There are no reports on the direct catalytic polymerization of $CO_2$ and diols, and the present work is the first example using $CeO_2$ catalyst and 2-cyanopyridine promotor, providing the alternating cooligomers in high diol-based yield (up to 99%) and selectivity (up to >99%). This combination of $CeO_2$ and 2-cyanopyridine is also effective to various diols including linear C4-C10 $\alpha,\omega$-diols to give high yields of the corresponding cooligomers, which cannot be synthesized by conventional methods such as copolymerization of $CO_2$ and cyclic ethers and ring-opening polymerization of cyclic carbonates.

# Catalytic copolymerization of $CO_2$ and diols using $CeO_2$ and 2-cyanopyridine

Keiichi Tomishige
*Department of Applied Chemistry, School of Engineering, Tohoku University, Aoba 6-6-07, Aramaki, Aoba-ku, Sendai 980-8579, Japan*
E mail: tomi@erec.che.tohoku.ac.jp

**Abstract**

Direct and catalytic polymerization of $CO_2$ and diols to polycarbonates is a promising technology as a simple and environmental-benign method. Generally, the synthesis of organic carbonates from alcohols and $CO_2$ needs selective catalysts and effective dehydrating agents because of the serious equilibrium limitation. One of the conventional synthesis methods of organic carbonates is the reaction of alcohols with phosgene. However, the problems of this phosgene method are the usage of poisonous phosgene and large amount of salt byproduct for the neutralization. Therefore the impact of the present work is related to the substitute for phosgene with $CO_2$. There are no reports on the direct catalytic polymerization of $CO_2$ and diols, and the present work is the first example using $CeO_2$ catalyst and 2-cyanopyridine promotor, providing the alternating cooligomers in high diol-based yield (up to 99%) and selectivity (up to >99%). This combination of $CeO_2$ and 2-cyanopyridine is also effective to various diols including linear C4-C10 α,ω-diols to give high yields of the corresponding cooligomers, which cannot be synthesized by conventional methods such as copolymerization of $CO_2$ and cyclic ethers and ring-opening polymerization of cyclic carbonates.

## 1. Introduction

Conversion of $CO_2$ into useful compounds can contribute to the construction of future sustainable society. Figure 1 illustrates $CO_2$ conversion technology, which can be divided to two categories: reductive conversion and non-reductive conversion.

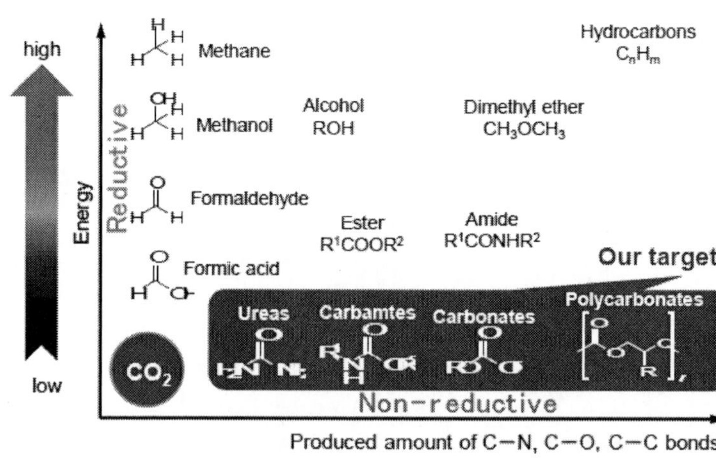


Figure 1. $CO_2$ conversion technology: reductive and non-reductive conversion

The reductive conversion includes the production of hydrocarbons and oxygenates from $CO_2$ by the reduction with $H_2$ and artificial photosynthesis. On the other hand, the compounds such as ureas, carbamates, and carbonates are also useful, and the oxidation state of the carbon atom in these compounds are the same as that in $CO_2$ molecule, which is regarded as non-reductive $CO_2$ conversion. Our target is the development of catalytic reaction systems for the non-reductive $CO_2$ conversion. Figure 2 shows the impact of the non-reductive $CO_2$ conversion, which is related to the substitution for phosgene with $CO_2$. Dimethyl carbonate (DMC) is an example of phosgene-derived chemicals, which is produced from phosgene and methanol by the reaction of $2CH_3OH+COCl_2+2NaOH\rightarrow(CH_3O)_2CO+2NaCl+H_2O$. The problem of this method is the usage of poisonous phosgene and the formation of large amount of salt for the neutralization, and this is not so suitable in terms of green chemistry. In contrast, the synthesis of DMC from methanol and $CO_2$ ($2CH_3OH+CO_2\rightarrow(CH_3O)_2CO+H_2O$) is more suitable than the phosgene method. This DMC synthesis should be promoted by effective catalysts because the reactivity of $CO_2$ is much lower than that of phosgene. Effective catalysts for this reaction are very limited, however, our group found that $ZrO_2$ [1, 2], $H_3PO_4/ZrO_2$ [3, 4], $CeO_2$-$ZrO_2$ [5, 6], and $CeO_2$ [7]. Another serious problem is the equilibrium limitation. Figure 3 shows the reaction time dependence of DMC synthesis from methanol and $CO_2$ using different catalyst amounts of $CeO_2$ [7]. Amount of DMC was saturated at the similar level over 10 and 100 mg $CeO_2$ catalyst, which indicates that this DMC amount is that at the equilibrium of the reaction of methanol and $CO_2$ to DMC. An important point is that DMC amount at the equilibrium is about 0.7 mmol, which is much smaller

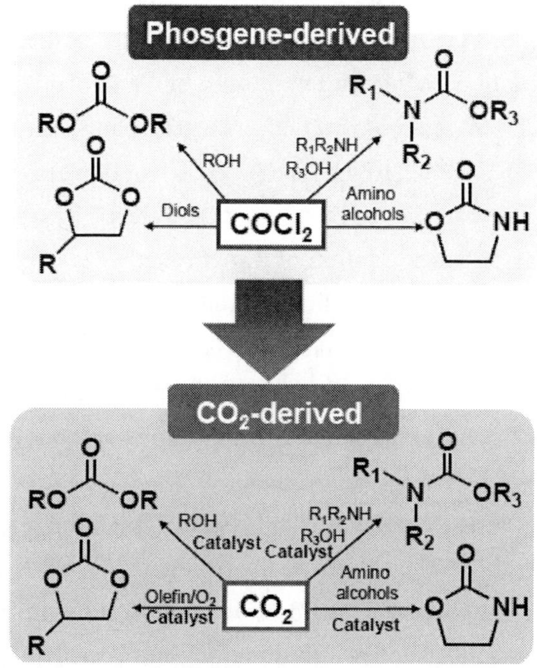


Figure 2. Substitution of phosgene with $CO_2$ by the non-reductive conversion of $CO_2$

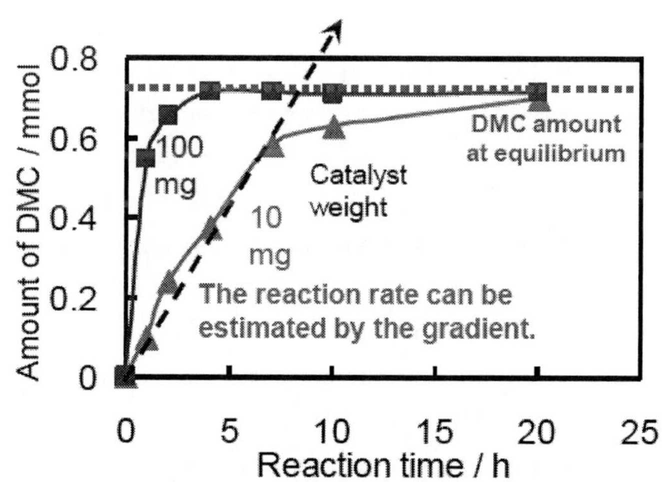


Figure 3. Reaction time dependence of DMC synthesis from methanol and $CO_2$ using a different amount of $CeO_2$ catalyst (10 and 100 mg).
Reaction conditions: $CH_3OH:CO_2$=192:200 mmol 403 K. Catalyst: $CeO_2$-HS calcined at 873 K.

than methanol and $CO_2$ amount in this system. This means very low equilibrium yield (below 1%) of DMC from methanol and $CO_2$. Another important point is the selectivity of DMC formation. When larger amount of $CeO_2$ (100 mg) was test at long reaction time (24 h), where the DMC amount reached the equilibrium level, the possible byproduct dimethyl ether ($2CH_3OH \rightarrow CH_3OCH_3+H_2O$) was not detected at all. The equilibrium level of DME is much higher than that of DMC, and the selectivity of DMC formation is almost 100% over $CeO_2$ catalyst. No formation of dimethyl ether is very important for the formation of DMC because water derived from dimethyl ether with higher equilibrium level than DMC inhibits the formation of DMC. It should be noted that the problem of the equilibrium in the reaction of $CO_2$ with alcohols to corresponding organic carbonates is much more serious than that in the production of carbamates from amines+alcohols+$CO_2$ [8-10] and ureas from amines+$CO_2$ [10-12].

In order to enhance the yield of DMC, the reaction equation of DMC from methanol and $CO_2$ should be shifted to the product side, for example, by the removal of $H_2O$ from the reaction system. One possible method is the utilization of the hydration reaction of organic compounds such as 2,2-dimethoxypropane [6, 13], etc. Among tested organic compounds, we found that acetonitrile hydration to acetoamide ($CH_3CN+H_2O \rightarrow CH_3CONH_2$) was very effective to the enhancement of the DMC yield, and the highest methanol-based DMC yield was 9% [14, 15]. After that, we also found that benzonitrile hydration to benzamide ($C_6H_5\text{-}CN+H_2O \rightarrow C_6H_5\text{-}CONH_2$) was more effective, however, the DMC yield was saturated to be 47% [16]. The saturation of the DMC yield is due to the catalyst poison of benzamide. During these works, the group of Prof. Satsuma and Prof. Shimizu reported that $CeO_2$ catalyzed the hydration of 2-cyanopyridine and the reaction rate on $CeO_2$ was much higher than that on various typical oxide catalysts [17]. Based on this report, we tested the effect of 2-cyanopyridine in combination with the synthesis of DMC from methanol and $CO_2$, giving very high yield of DMC [18, 19]. Figure 4 shows the comparison in the reaction time dependence between 2-cyanopyridine and benzonitrile in combination with DMC synthesis from methanol and $CO_2$ [16, 18]. In the case of 2-cyanopyridine, methanol was converted almost completely and the yield of DMC was not suppressed at all, indicating that 2-picolinamide is not a catalyst poison and the nitrogen atom in the pyridine ring plays an essential role on weakening the interaction between the amide group and the surface of $CeO_2$. After the optimization, the highest methanol-based DMC yield reached 94% at the reaction conditions where the molar ratio of $CH_3OH$ to 2-cyanopyridine is 2 (stoichiometric ratio: $2CH_3OH+CO_2+C_5NH_4\text{-}CN \rightarrow (CH_3O)_2CO+ C_5NH_4\text{-}CONH_2$). Another important finding is that the presence of 2-cyanopyridine enhanced the formation rate of DMC drastically [19]. We also carried out the characterization of the interaction between $CeO_2$ surface and 2-cyanopyridine, DFT calculation, and the activity test of base-catalyzed reactions, and it is proposed that the interaction of 2-cyanopyridine with the surface of $CeO_2$ can form the strong basic site, which is connected to high catalytic activity of the reactions including the DMC synthesis from methanol and $CO_2$ [20-22].

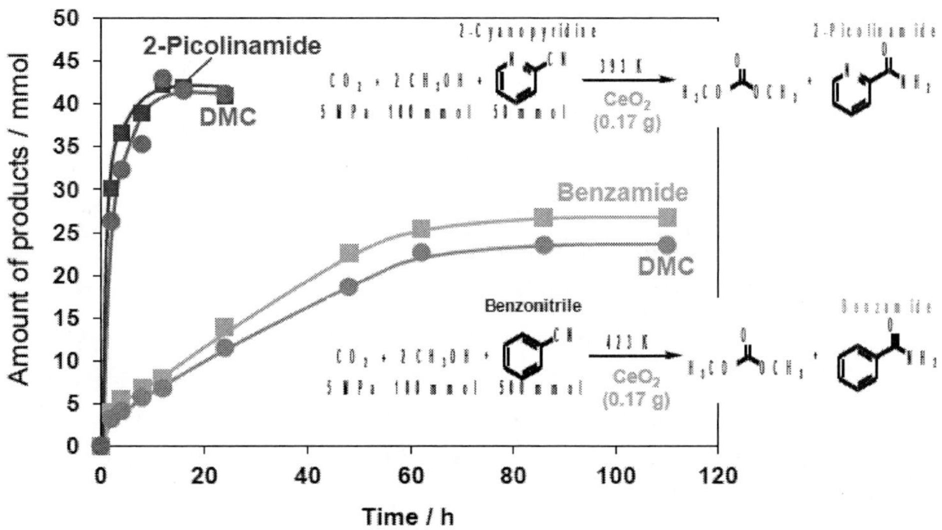


Figure 4. Comparison in the reaction time dependence between 2-cyanopyridine and benzonitrile in the combination with DMC synthesis from methanol and $CO_2$.

Based on these results, we applied the reaction system using 2-cyanopyridine to the reaction of diols with $CO_2$. We also reported that $CeO_2$ also catalyze the synthesis of 5-membered cyclic carbonates such as propylene carbonate and ethylene carbonate from the reaction of 1,2-propanediol and ethylene glycol with $CO_2$, where the serious equilibrium limitation without $H_2O$ removal [23, 24]. Here, we report that the high yield synthesis of cyclic carbonates and polycarbonates from the reaction of diols and $CO_2$ using 2-cyanopyridine as a dehydrating agent by the hydration of nitrile group.

## 2. Experimental [25, 26]

Preparation of $CeO_2$ catalyst was carried out by calcining $CeO_2$-HS (Daiichi Kigenso, Japan. The purity of $CeO_2$: 99.97%) for 3 h in air at 873 K. The specific surface area (BET method) of the $CeO_2$ was 84 $m^2/g$.

All the reactions were carried out in an autoclave reactor with inner volume of 190 mL. A typical procedure of the reaction of diols and $CO_2$ with 2-cyanopyridine to corresponding cyclic carbonates is as follows: 2 mmol of the $CeO_2$ catalyst (0.34 g), 10 mmol of diol, and 100 mmol of 2-cyanopyridine were put into the autoclave together with a spinner, and then the reactor was purged and pressurized with $CO_2$. Then the reactor was heated to the reaction temperature. After the reaction time, the reactor was cooled to room temperature, and 30 mL of ethanol and 0.2 mL of 1-hexanol were added to the liquid phase as a solvent and an internal standard substance for a quantitative analysis, respectively. Products in liquid and gas phases were analyzed by using a gas chromatograph (GC) equipped with an FID and GC/MS.

The standard procedure of direct polymerization of $CO_2$ and 1,4-butanediol using the combination catalyst of $CeO_2$ and 2-cyanopyridine is as follows: $CeO_2$ (0.17 g, 1 mmol), 1,4-butanediol 0.90 g (10 mmol) and 2-cyanopyridine 10.4 g (100 mmol) were put into the autoclave together with a spinner, and then the reactor was purged with 1 MPa $CO_2$ three times. The autoclave was pressurized with $CO_2$ to the desired pressure (typically 5.0 MPa) at room temperature, and then the autoclave was heated to 433 K, where the $CO_2$ pressure was about 12 MPa. The mixture was constantly stirred during the reaction. After the reaction time, the reactor was cooled in water bath to room temperature. Tetrahydrofuran (THF) (20 ml) was added to the liquid phase as a solvent, and 1-hexanol (~0.2 ml) was also added as an internal standard substance for a quantitative analysis. The reactor was washed with THF, and the liquids used in washing were added to the reaction mixture. Amount of 2-cyanopyridine and products from 2-cyanopyridine such as 2-picolinamide and 4-hydroxybutyl picolinate were analyzed by GC equipped with an FID. Since produced cooligomers are decomposed by heating, the amount of 1,4-butanediol was analyzed by HPLC equipped with an RI detector at the conditions of developing solvent $H_2O/CH_3OH = 70/30$, 0.5 ml/min, 313 K. Since produced cooligomers were precipitated by addition of the developing solvent (about 20-fold dilution), the precipitated cooligomers were removed by filtration before analyzing by HPLC. This filtration operation was conducted at least two times until the precipitation was not observed. The qualitative analysis of the products was conducted by GC-MS and NMR. The oligomerized products were analyzed by MALDI-TOF mass using dithranol and NaBr as a matrix and ionization agent, respectively, and size exclusion chromatography with a RI detector. The developing solvent is THF. The conversion and selectivity were calculated by the following equations (Eqs. 1-3).

Conversion/% = 100 × [1 − (Amount of diol after reaction(mmol))/(Amount of diol before reaction(mmol))] (1)

Selectivity/% = 100 × [(Amount of product(mmol) (number of diol units in product) /(amount of reacted diol(mmol))] (2)

Yield/% = Conversion(%) × selectivity(%)/100 (3)

**3. Results and discussion**

3.1. Synthesis of cyclic carbonates from diols and $CO_2$ [25]

Like the case of DMC synthesis from methanol and $CO_2$, the reaction of diols with $CO_2$ to corresponding carbonates is also limited seriously by the equilibrium, however, the combination of the hydration of 2-cyanopyridine enhanced the yield of cyclic carbonates drastically. Figure 5 shows the results of the synthesis of cyclic carbonates from diols and $CO_2$ catalyzed by $CeO_2$ in the presence of 2-cyanopyidine. Various kinds of 5-membered-ring cyclic carbonates were obtained in high yield from vicinal diols and $CO_2$. It is thought that

5-membered-ring carbonates can be synthesized easily by the reaction of $CO_2$ with corresponding olefin oxides. In contrast, it has been known that the synthesis of 6-membered-ring carbonates from the reaction of $CO_2$ with corresponding oxetanes is not so easy. Therefore, we tested the synthesis of 6-membered-ring carbonates from 1,3-diols and $CO_2$, indicating that the combination of $CeO_2$+2-cyanopyridine was very effective to the synthesis of 6-membered-ring cyclic carbonates in high yield.

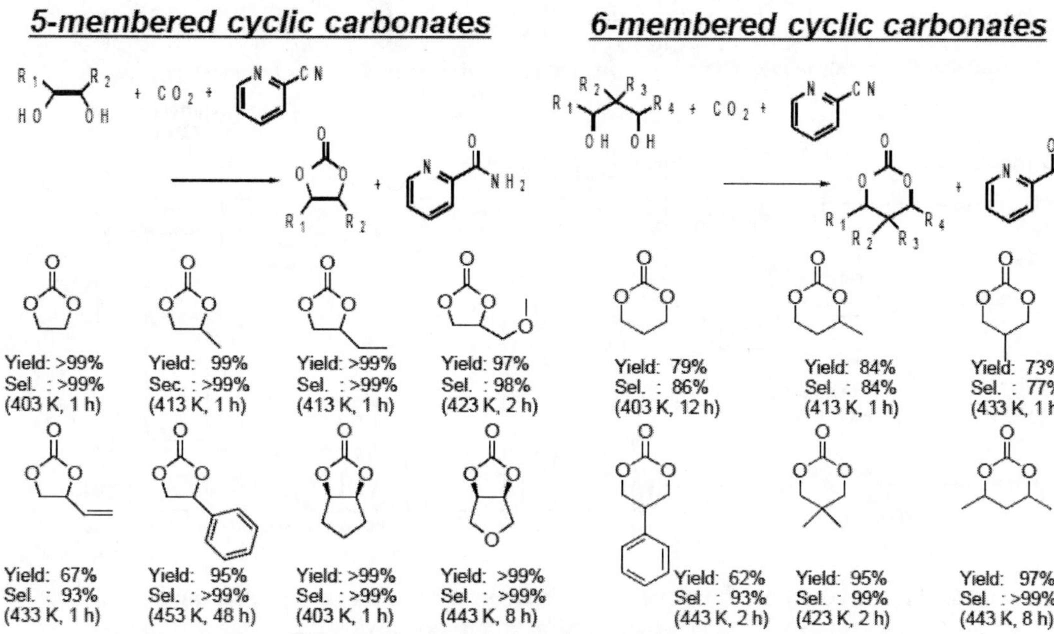


Figure 5. Synthesis of cyclic carbonates from diols and $CO_2$ catalyzed by $CeO_2$ in the presence of 2-cyanopyidine. Reaction conditions: $CeO_2$ 0.34 g, diol : 2-cyanopyridine = 10 mmol : 100 mmol, $CO_2$ 5 MPa.

Another interesting result is the reaction time dependence in the reaction of 1,3-propanediol and $CO_2$ catalyzed by $CeO_2$ in the presence of 2-cyanopyridine (Figure 6 (a)). At shorter reaction time, the selectivity to trimethylene carbonate was high, however, it is clearly lower than 100% and the selectivity decreased with the reaction time. In fact, other products were not detected by the GC analysis. Therefore, the products were also analyzed by HPLC and the polymerized products were detected. This is also supported by the result of MALDI-TOF-MS analysis, suggesting the formation of an alternating copolymer of $CO_2$ and 1,3-propanediol. Stability and reactivity of cyclic carbonates are strongly dependent on the number of atoms giving the ring structure. Five membered-ring cyclic carbonates have high stability and low reactivity. Six membered-ring cyclic carbonates have lower stability and higher reactivity than 5-membered ring ones. Seven and more membered-ring cyclic carbonates have much lower stability and higher reactivity. Therefore, 5-membered-ring cyclic carbonates formation is easier than 6-membered-ring carbonates, which can explain the lower selectivity to

6-membered-ring carbonates. At the same time, longer diols than 1,3-propanediol will give the polycarbonates with higher selectivity because of low stability and high reactivity of the corresponding cyclic carbonates. It should be noted that seven and more membered-ring carbonates is difficult to synthesize and use as a monomer for the production of polymer because of low stability.

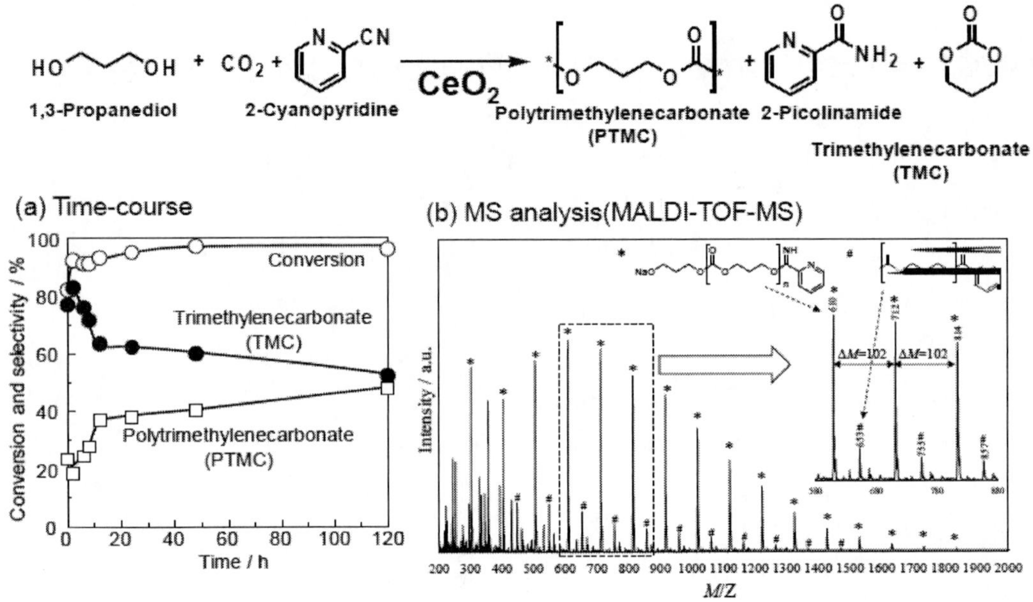


Figure 6. Synthesis of cyclic carbonate and polycarbonate from 1,3-propanediol and $CO_2$ catalyzed by $CeO_2$ in the presence of 2-cyanopyidine.
(a) Reaction time dependence, (b) MALDI-TOF-MS of the produced polycarbonate.
Reaction conditions: $CeO_2$ 0.17 g, 1,3-propanediol : 2-cyanopyridine = 10 mmol : 100 mmol, $CO_2$ 5 MPa.

## 3.2. Synthesis of polycarbonates from diols and $CO_2$ [26]

Figure 7 shows the results of the reaction of 1,4-butanediol, $CO_2$, and 2-cyanopyridine catalyzed by $CeO_2$. The conversion of 1,4-butanediol was high from the initial state and the formation of cyclic carbonates is not observed, indicating that the oligomer is formed selectively. According to the result of the molecular weight (Figure 7 (b)), the polymerization reaction proceeded until 10 h of reaction time. After the maximum, the molecular weight decreased slightly by the back-biting mechanism. On the basis of the reaction time dependence that conversion of 1,4-butanediol was high even at the initial state, it is proposed that the dimerization of 1,4-butanediol proceeds at first, and the subsequent oligomerization proceeds (Figure 8). Another important point is that the oligomerization can proceed without the formation of 7-membered cyclic carbonate and this means the catalytic and direct copolymerization of diols and $CO_2$.

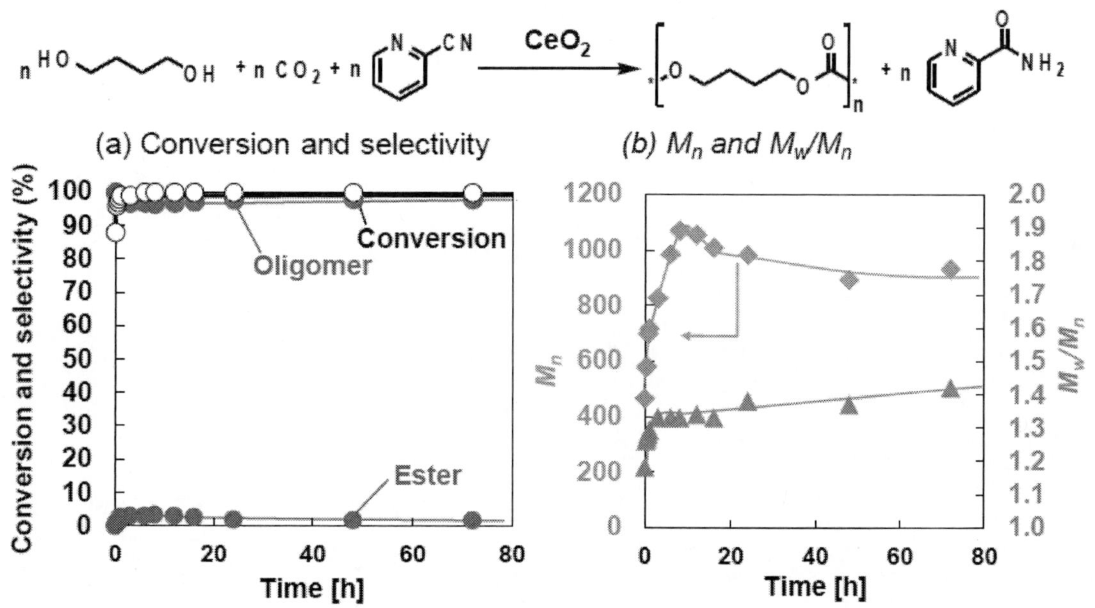


Figure 7. Results of the reaction of 1,4-butanediol, $CO_2$, and 2-cyanopyridine catalyzed by $CeO_2$ (a) Reaction time dependence, (b) $M_n/M_w$ ($M_n$: number-average molecular weight, polystyrene equivalent molar mass (SEC); $M_w$: weight-average molecular weight).
Reaction conditions: $CeO_2$ 0.17 g, 1,4-butanediol 10 mmol, 2-cyanopyridine 100 mmol, $CO_2$ 5 MPa, 403 K.

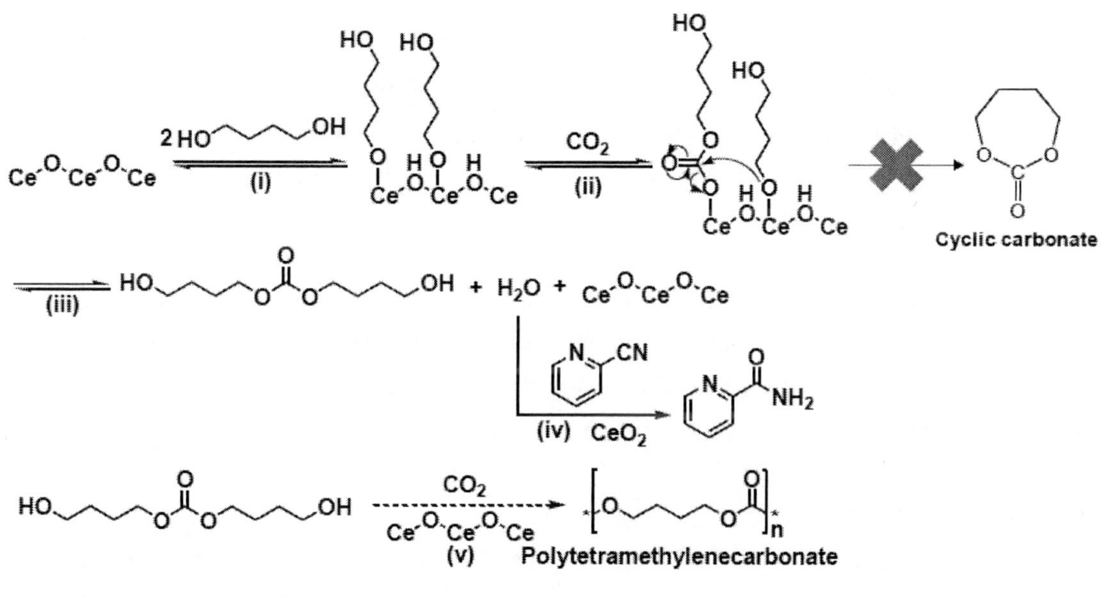


Figure 8. Proposed reaction scheme of the reaction of 1,4-butanediol, $CO_2$, and 2-cyanopyridine catalyzed by $CeO_2$

Table 1. Result of the reaction of $CO_2$ with various diols catalyzed by $CeO_2$ in the presence of 2-cyanopyridine

| Entry | Diol [mmol] | Conv. [%] | Sel. [%] | $M_n$ | $M_w/M_n$ | The number of the repeating unit |
|---|---|---|---|---|---|---|
| 1[a] | 1,4-Butanediol | >99 | 98 | 1070 | 1.33 | 9.2 |
| 2 | 1,5-Pentanediol | 98 | 99 | 930 | 1.34 | 7.1 |
| 3 | 1,6-Hexanediol | 99 | 99 | 1080 | 1.31 | 7.5 |
| 4 | 1,8-Octanediol | 99 | 97 | 1200 | 1.33 | 7.0 |
| 5 | 1,10-Decanediol | 93 | 98 | 1650 | 1.26 | 8.3 |
| 6 | | 42 | 98 | 510 | 1.04 | 3.0 |
| 7 | | 54 | 82 | 590 | 1.10 | 3.6 |
| 8 | 1,5-Hexanediol | 93 | 99 | 530 | 1.10 | 3.7 |
| 9 | 2,5-Hexanediol | 6 | 60 | 450 | 1.01 | 3.1 |
| 10 | 2,5-Dimethyl-2,5-hexanediol | <1 | <1 | - | - | - |

Reaction conditions: $CeO_2$ 0.17 g, diol 10 mmol, 2-cyanopyridine 100 mmol, $CO_2$ 5 MPa, 403 K, 24 h. [a] 8 h.

Table 1 lists the result of the reaction of $CO_2$ with various diols catalyzed by $CeO_2$ in the presence of 2-cyanopyridine. The $\alpha,\omega$-diols (primary-primary) such as 1,4-butanediol, 1,5-pentanediol, 1,6-hexanediol, 1,8-octanediol, and 1,10-decanediol showed high conversion and selectivity to polycarbonate, although the polymerization degree is not high. Even in the presence of cyclohexane and benzene ring, primary-primary diols can give the oligomerized product, although the reactivity of such substrates is not high. Primary-secondary diol such as 1,5-hexanediol can give the oligomerized product, however, the reaction of secondary-secondary and tertiary-tertiary diol is difficult to proceed, and hence these diols are not suitable substrates for $CeO_2$+2-cyanopyridine system.

## 5. Conclusions

Direct polymerization of $CO_2$ and diols is promising as a simple and environmental-benign method in place of conventional processes using high-cost and/or hazardous reagents such as phosgene, carbon monoxide and epoxides, however, there are no reports on the direct method due to the inertness of $CO_2$ and severe equilibrium limitation of the reaction. The present work is the first example of the catalytic and direct copolymerization of $CO_2$ and diols using $CeO_2$ catalyst and 2-cyanopyridine promotor, providing polycarbonates in high diol-based yield (up to 99%) and selectivity (up to >99%). This catalyst system including

$CeO_2$-catalyzed carboxylation of diols and hydration of 2-cyanopyridine is applicable to various diols including linear C4-C10 α,ω-diols to provide high yields of the corresponding cooligomers, which cannot be obtained by well-known methods such as copolymerization of $CO_2$ and cyclic ethers and ring-opening polymerization of cyclic carbonates. The proposed process provides us a facile synthesis method for versatile polycarbonates from various diols and $CO_2$ owing to simplicity of diols modification.

## 6. References

[1] K. Tomishige, T. Sakaihori, Y. Ikeda, K. Fujimoto, Catal. Lett., 58, 225 (1999).
[2] K. Tomishige, Y. Ikeda, T. Sakaihori, K. Fujimoto, J. Catal., 192, 355 (2000).
[3] Y. Ikeda, T. Sakaihori, K. Tomishige, K. Fujimoto, Catal. Lett., 66, 59 (2000).
[4] Y. Ikeda, M. Asadullah, K. Fujimoto, K. Tomishige, J. Phys. Chem. B, 105, 1065 (2001)
[5] K. Tomishige, Y. Furusawa, Y. Ikeda, M. Asadullah, K. Fujimoto, Catal. Lett., 76, 71 (2001)
[6] K. Tomishige, K. Kunimori, Appl. Catal. A, 237, 103 (2002).
[7] Y. Yoshida, Y. Arai, S. Kado, K. Kunimori, K. Tomishige, Catal. Today, 115, 95 (2006).
[8] M. Honda, S. Sonehara, H. Yasuda, Y. Nakagawa, K. Tomishige, Green Chem., 13, 3406 (2011).
[9] M. Tamura, K. Noro, M. Honda, Y. Nakagawa, K. Tomishige, Green Chem., 15, 1567 (2013).
[10] M. Tamura, M. Honda, Y. Nakagawa, K. Tomishige J. Chem. Technol. Biotechnol., 89, 19 (2014).
[11] M. Tamura, K. Ito, Y. Nakagawa, K. Tomishige, J. Catal., 343, 75 (2016).
[12] M. Tamura, M. Honda, K. Noro, Y. Nakagawa, K. Tomishige, J. Catal., 305, 191 (2013).
[13] M. Honda, M. Tamura, Y. Nakagawa, K. Tomishige, Catal. Sci. Technol., 4, 2830 (2014).
[14] M. Honda, A. Suzuki, N. Begum, K. Fujimoto, K. Suzuki, K. Tomishige, Chem. Commun., 4596 (2009).
[15] M. Honda, S. Kuno, N. Begum, K. Fujimoto, K. Suzuki, Y. Nakagawa, K. Tomishige, Appl. Catal. A, 384, 165 (2010).
[16] M. Honda, S. Kuno, S. Sonehara, K. Fujimoto, K. Suzuki, Y. Nakagawa, K. Tomishige, ChemCatChem, 3, 365 (2011).
[17] M. Tamura, H. Wakasugi, K. Shimizu, A. Satsuma, Chem. Eur. J., 17, 11428 (2011).
[18] M. Honda, M. Tamura, Y. Nakagawa, S. Sonehara, K. Suzuki, K. Fujimoto, K. Tomishige, ChemSusChem, 6, 1341 (2013).
[19] M. Honda, M. Tamura, Y. Nakagawa, K. Nakao, K. Suzuki, K. Tomishige, J. Catal., 318, 95 (2014).
[20] M. Tamura, R. Kishi, A. Nakayama, Y. Nakagawa, J. Hasegawa, K. Tomishige, J. Am.

Chem. Soc., 139, 11857 (2017).

[21] M. Tamura, R. Kishi, Y. Nakagawa, K. Tomishige, Nature Commun., 6, 8580 (2015)

[22] M. Tamura, K. Sawabe, K. Tomishige, A. Satsuma, K. Shimizu, ACS Catal., 5, 20 (2015).

[23] K. Tomishige, H. Yasuda, Y. Yoshida, M. Nurunnabi, B. Li, K. Kunimori, Green Chem., 6, 206 (2004).

[24] K. Tomishige, H. Yasuda, Y. Yoshida, M. Nurunnabi, B. Li, K. Kunimori, Catal. Lett., 95, 45 (2004).

[25] M. Honda, M. Tamura, K. Nakao, K. Suzuki, Y. Nakagawa, K. Tomishige, ACS Catal., 4, 1893 (2014).

[26] M. Tamura, K. Ito, M. Honda, Y. Nakagawa, H. Sugimoto, K. Tomishige, Sci. Rep., 6, 24038 (2016).

Paper No. 13

# Toluene methylation to produce para-Xylene using ZSM-5 catalysts

***Arudra Palani***, *KFUPM-RI;* ***Ali Jahel,*** *Honeywell UOP*

Para-Xylene (PX) is an industrially important raw material employed in the large scale production of terephthalic acid (TPA) and its derivatives used to produce thermoplastic polyesters [1-4]. The global paraxylene market was valued at 34.74 billion (USD) in 2015 and it is projected to reach about 66.93 billion (USD) by 2022. Typically, *p*-xylene is produced in a petroleum refinery via catalytic reforming of naphtha, toluene disproportionation (TDP) and transalkylation of higher aromatics [5-6]. However, the concentration of *p*-xylene in xylenes produced by these processes is low, which requires high energy cost separation process [7]. To avoid separation process, selective production of *p*-xylene has more attention and challenging one. Alkylation of toluene with methanol is a promising and competitive pathway for highly selective production of *p*-xylene [8]. ZSM-5 zeolites have been well recognized as shape selective catalyst. The shape selectivity arises due to the straight and zigzag channels with estimated pore size opening of 0.55-0.57 nm [9]. These channels produce *p*-xylene preferentially as it diffuses faster than the other two isomers. However, the non-selective acid sites situated on the surface, allow *p*-xylene to isomerize leading to the thermodynamic equilibrium [10]. To enhance *para*-selectivity for toluene alkylation with methanol, various methods have been developed to modify the ZSM-5 catalysts.

In our present study, large crystal high silica MFI (Si/Al molar ratio 2000) catalyst was synthesized using fluoride method. About 2 grams of catalyst was mixed with 6 grams of SiC and the combined material was loaded in a 19 mm (0.75 inch) ID stainless steel tubular reactor. Before catalytic run, the catalyst was activated in a hydrogen stream at 500°C for 1 hour, and then a mixture of the feed (toluene and methanol) and hydrogen was passed through the catalyst bed. The total reactor effluent vapor composition was analyzed through an on-line GC using an FID (Innovax column) as well as TCD (GS Q column). Figure 1.shows the catalytic activity of MFI-2000 tested at a reactor pressure of 10 Barg (145 psig) up to 22 hours on stream. At 500°C, the toluene conversion was about 20% with Px/X = 67 and total xylenes yield of 14.9%. MFI-2000 catalyst showed stable activity with higher para-xylene selectivity with lower benzene to xylene ratio. These clearly indicate that MFI-2000 was a suitable candidate for the selective production of para-xylene.

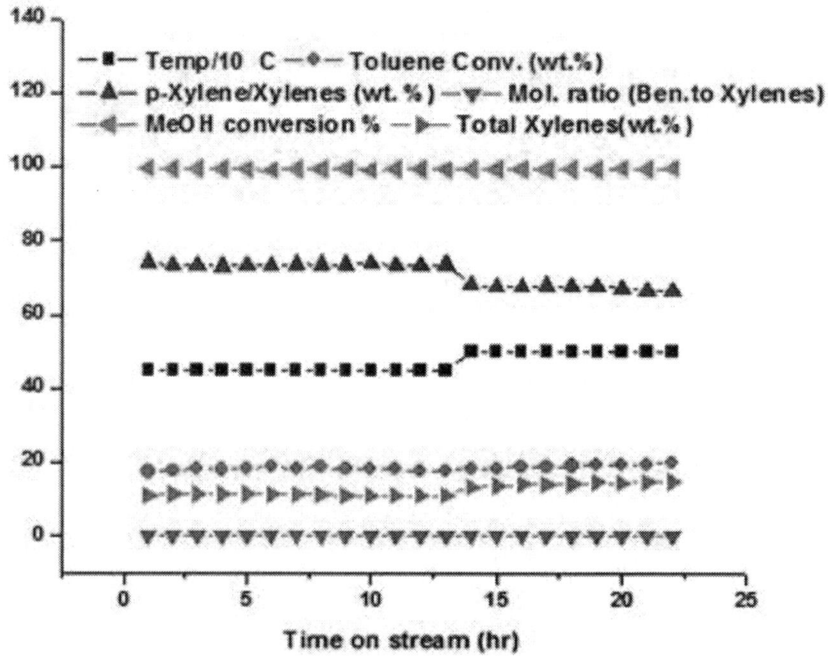


Figure 1. Catalytic activity of large silica MFI-2000 at a reactor pressure of 10 Barg (145 psig), WHSV=2; H2: HC= 4.

## References


1. Vermeiren, W.; Gilson, J.-P. Top. Catal. 2009, 52, 1131–1161.
2. T.-C. Tsai, S.-B. Liu, I. Wang, Appl. Catal. A. 1999, 181, 355–398.
3. H.-K. Min, S.H. Cha, S.B. Hong, ACS Catal. 2012, 2, 971– 981.
4. N.M. Tukur, S. Al-Khattaf, Chem. Eng. J. 2011, 166, 348–357.
5. Jagannath Das, Yajnavalkya S. Bhat and Anand B. Halgeri, Catalyst Letters. 1994, 23, 161—168.
6. Ikai Wang, Tseng Chang Tsai, Sheng Tai Huang, Ind. Eng. Chem. Res., 1990, 29, 2005–2012
7. Muhammad Tahir Ashraf, Rachid Chebbi, Naif A. Darwish, Ind. Eng. Chem. Res., 2013, 52, 13730–13737.
8. Ghosh, A. K., and Harvey, P. Preparation of Alkyl-aromatic Products. 2006, U.S. Patent 7105713B2.
9. Odedairo T., Balasamy R. J., Al-khattaf S, Ind. Eng. Chem. Res., 2011, 50, 3169–3183.
10. S. Zheng, A. Jentys, J.A. Lercher, J. Catal. 2003, 219, 310–319.

Paper No. 14

# Metathesis of 2-pentene over Mo and W Supported Mesoporous Molecular Sieves

***M. Naseem Akhtar***
*Center for Refining & Petrochemicals, KFUPM-RI, Dhahran*

Molybdenum and tungsten oxides were supported on silica, MCM-22, MCM-41 and SBA-15. XRD and N2 adsorption-desorption revealed that architecture and textural character of supports were preserved. The catalysts were investigated in transformation of 2-pentene at different reaction temperatures. MoO3/MCM-22 exhibited highest conversions with isomerization and cracking as major reactions. MoO2(acac)2, MoO3 and WO3 supported on MCM-41 and SBA-15 showed metathesis reaction of 2-C5= producing propylene, C4= and C6+= as major products. Catalysts based on MCM-41 exhibited higher activity and stability than SBA-15. Addition of ethylene to 2-C5= increased selectivity to propylene due to metathesis of ethylene with 2-pentene.

Paper No. 15

## Carbon Dioxide Fixations with Homogeneous Catalysis

***Yasushi Tsuji***
*Kyoto University*

Carbon dioxide ($CO_2$) is a readily available and renewable chemical feedstock. However, the thermodynamic considerations of $CO_2$ limit its widespread use in chemical reactions. We have been developing the fixations of $CO_2$ in the presence of a homogeneous catalyst by utilizing carbon-carbon bond-forming reactions. A Ni complex catalyzes the carboxylation of less reactive aryl chlorides with $CO_2$. Double carboxylation of internal alkynes, in which two $CO_2$ molecules are fixed into alkynes, is realized under 1 atm of $CO_2$ in the presence of a Ni catalyst combined with a suitable reducing agents such as Mn or Zn powders. When a Cu complex is used as a catalyst under $CO_2$ atmosphere, hydrocarboxylation and silacarboxylation of alkynes proceed efficiently employing hydrosilanes and silylboranes, respectively.

# Carbon Dioxide Fixations with Homogeneous Catalysis

Yasushi Tsuji
ytsuji@scl.kyoto-u.ac.jp
*Department of Energy and Hydrocarbon Chemistry*
*Kyoto University*
*Kyoto 615-8510, Japan*

**Abstract**

Carbon dioxide ($CO_2$) is a readily available and renewable chemical feedstock; however, the thermodynamic considerations of $CO_2$ limit its widespread use in chemical reactions. This article summarizes the transformation of $CO_2$ via carbon-carbon bond-forming reactions. Ni complexes catalyzed the carboxylation of less reactive aryl chlorides and double carboxylation of internal alkynes in the presence of suitable reducing agents such as Mn or Zn powders under 1 atm of $CO_2$. When Cu complexes were used as catalysts under $CO_2$ atmosphere, hydrocarboxylation and silacarboxylation of alkynes proceeded efficiently using hydrosilanes and silylboranes, respectively.

## Introduction

Carbon dioxide ($CO_2$) is an ideal carbon source owing to its abundance, low cost, nontoxicity, and good potential as a renewable source.[1,2] However, it is not easy to activate $CO_2$ because of its thermodynamic and kinetic stability. Therefore, homogeneous transition-metal catalysts play an important role in developing useful transformations by utilizing this attractive and environment-friendly raw material.

Three important methodologies are available to achieve the transition-metal-catalyzed fixation of $CO_2$. The carbon–hydrogen bond-forming reactions of $CO_2$ is important for the syntheses of C1 chemicals such as formic acid derivatives, methanol, and methane.[3-6] Organic carbonates can be synthesized by the carbon–oxygen bond-forming reactions of $CO_2$ with epoxides.[7-12] The carbon–carbon (C–C) bond-forming reactions using $CO_2$ are straightforward methods for synthesizing diverse carboxylic acids and their derivatives.[12-20]

In the last decade, much attention has been paid to develop the catalytic fixation of $CO_2$ via C–C bond formation. The most important factor in catalytic fixation of $CO_2$ is the formation of M–C bond where M is a transition metal used as the catalyst. To date, three main elemental steps have been reported for the catalytic fixation of $CO_2$ (Scheme 1). The first step is a substitution- type reaction (Scheme 1a). M–C bond can be formed catalytically by the oxidative addition of organic halides (C–X) or the transmetalation with organometallic reagents. Then, the insertion of $CO_2$ into the M–C bond affords the corresponding carboxylic acids. In the oxidative addition, the catalytic carboxylation of organic halides can be achieved using a suitable reducing agent.[21-28] The carboxylation reactions of aryl and alkylzinc reagents[29-31] as well as aryl,[32-34] allyl,[35] and alkylboron[36,37] compounds with $CO_2$ have been widely studied, because various functional groups that are not compatible with Grignard reagents are tolerant in these reactions. The second step is the addition of the catalytically generated organometallic species to C–C multiple bonds (Scheme 1b). In particular, the hydrocarboxylation reaction including the addition of M–H species across C–C multiple bonds such as alkynes,[39,40] alkenes,[41] 1,2-dienes (allenes),[42,43] and 1,3-dienes[44] followed by trapping $CO_2$ is very promising. The third step is the oxidative cyclization of a low-valent transition metal such as Ni(0) with C–C unsaturated compounds (alkenes or alkynes), and $CO_2$, affording the corresponding metallacycle intermediate (Scheme 1c). Interesting catalytic transformations have been often reported using these elemental step.[45-51]

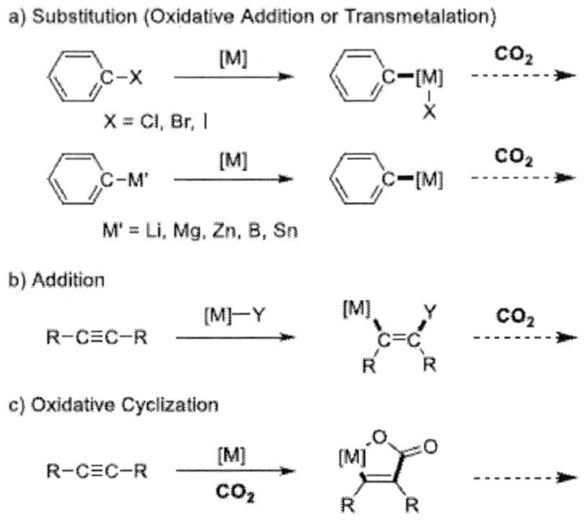


**Scheme 1.** Elemental steps for the transition-metal catalyzed fixation of $CO_2$ via C-C bond-forming reactions: a) Substitution reaction, b) addition reaction, and c) oxidative cyclization

Herein, we describe Ni- and Cu-catalyzed fixations of $CO_2$ which were established in our laboratory. The most important factor for the four successful transformations is the catalytic formation of M–C bonds (M = Ni or Cu) and regeneration of catalytically active species using suitable reducing reagents. The substrate scope and reaction mechanism of these reactions are discussed.

## Experimental

Unless otherwise noted, reactions were performed under an argon atmosphere using heat-gun-dried glassware on a dual-manifold Schlenk line. Most chemicals obtained from commercial suppliers were used without further purification.

## Results and discussion

### Carboxylation of Aryl Chlorides

The carboxylation of organic halides is one of the most straightforward methods to synthesize various carboxylic acids. In 2009, the first catalytic carboxylation of aryl bromides using $CO_2$ and highly reactive $ZnEt_2$ as the reducing agent was reported.[28] We reported a much more efficient catalytic reaction, in which a less noble metal Ni catalyst was highly active in the carboxylation of aryl chlorides under 1 atm pressure of $CO_2$ at room temperature using easy-to-handle Mn powder as the reducing agent.[27] The reaction of 1-butyl-4-chlorobenzene (**1a**) was carried out using a mixture of $NiCl_2(PPh_3)_2$ (5.0 mol%) and $PPh_3$ (10 mol%) as the catalyst and Mn powder (3.0 equiv) as a reducing agent in the presence of tetraethylammonium iodide ($Et_4NI$, 10 mol%) in 1,3-dimethyl-2-imidazolidinone (DMI) at 25 °C under 1 atm pressure of $CO_2$ (Scheme 2). After the derivatization to the corresponding methyl ester with trimethylsilyldiazomethane, methyl 4-butylbenzoate (**2a**-Me) was obtained in 95% GC yield. Without $PPh_3$, i.e., only $NiCl_2(PPh_3)_2$ as the catalyst, the yield of **2a**-Me reduced to 53%. In the absence of $NiCl_2(PPh_3)_2$, Mn powder or $Et_4NI$, the carboxylation reaction did not proceed at all, indicating that each component is essential for the reaction. Under the optimum reaction conditions, 4-butylbenzoic acid (**2a**) was isolated in 84% yield (Scheme 3). Aryl chlorides bearing both electron-rich and electron-deficient moieties afforded the corresponding carboxylic acids (**2b**–**2d**) in high yields. Ester and amide functional groups, which are not tolerated with organolithium or organomagnesium reagents, were well compatible under these reaction conditions (**2f**–**2h**). A boronic acid ester (**2i**) and a thiophene ring (**2j**) were also intact.

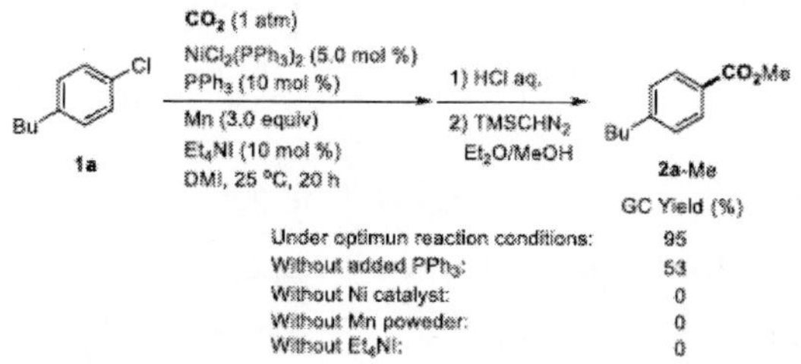


**Scheme 2.** Ni-catalyzed carboxylation of 4-butyl-1-chlorobenzene (**1a**)

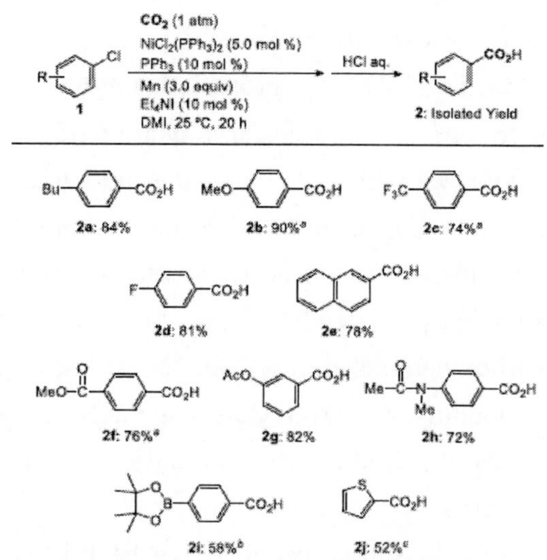


[a] $NiCl_2(P(p\text{-}MeOC_6H_4)_3)_2$ (0.025 mmol, 5 mol %), $P(p\text{-}MeOC_6H_4)_3$ (0.050 mmol, 10 mol %) at 40 °C, 24 h. [b] at 35 °C, 24 h. [c] at 35 °C, 30 h.

**Scheme 3.** Ni-catalyzed carboxylation of aryl chlorides (**1**)

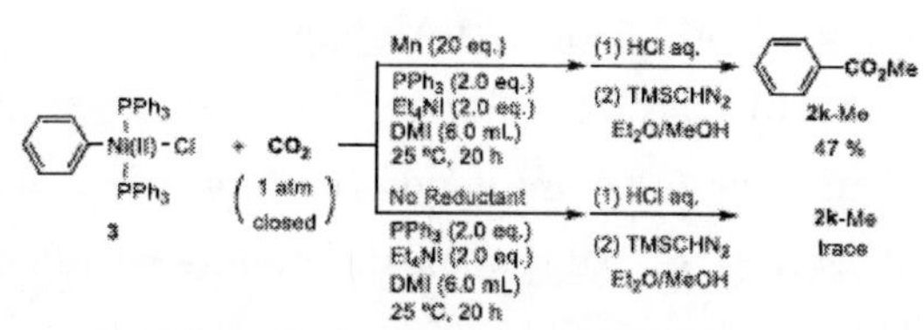


**Scheme 4.** Stoichiometric reactions relevant to mechanism employing $NiPhCl(PPh_3)_2$ (**3**)

To elucidate the catalytic mechanism, stoichiometric reactions were carried out using $NiPhCl(PPh_3)_2$ (**3**),[52] which was prepared by the oxidative addition of chlorobenzene to $Ni(PPh_3)_4$ (Scheme 4). In the presence of $CO_2$ (1 atm pressure), Mn powder, $Et_4NI$, and $PPh_3$,

**3** afforded the carboxylated product (**2k**-Me) in 47% yield after the derivatization to the corresponding methyl ester. However, upon the removal of either Mn powder or $Et_4NI$ from the reaction systems, only a trace amount of **2k**-Me was detected. Thus, both Mn and $Et_4NI$ are indispensable in the catalytic system and may reduce the Ni(II) to the corresponding Ni(I). The Ni(I) species should play an important role in the present catalytic carboxylation.

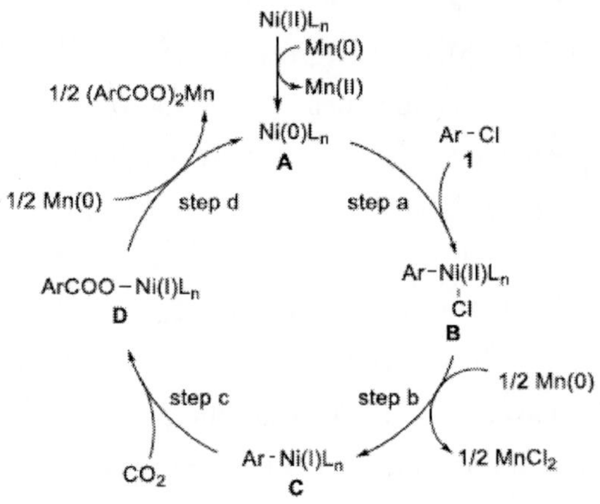


**Scheme 5.** Plausible reaction mechanism of Ni-catalyzed carboxylation of aryl chlorides (**1**)

Scheme 5 shows a possible catalytic cycle for the Ni-catalyzed carboxylation of aryl chlorides with $CO_2$. First, the oxidative addition of an aryl chloride (**1**) to Ni(0) (**A**) occurs, affording an aryl-Ni(II) intermediate (**B**) (step a). As suggested by the stoichiometric reaction in Scheme 4, Ni(II) would be reduced by the Mn/$Et_4NI$ system to afford Ni(I) intermediate (**C**) (step b).[53] Then, the more nucleophilic Ni(I) (**C**) reacts with $CO_2$, affording the carboxylatonickel intermediate (**D**) (step c). Finally, the reduction of **D** by Mn gives the corresponding Mn carboxylate, and the Ni(0) catalyst species (**A**) is regenerated (step d). The mechanism including the formation of the Ni(I) species is also supported by DFT caclulations.[54]

Using the similar protocol, we established the Co-catalyzed carboxylation of propargyl acetates.[55] With Mn under 1 atm pressure of $CO_2$ in the presence of a Co catalyst, diverse propargyl acetates were converted to the corresponding carboxylic acids in good to high yields.

Double Carboxylation of Alkynes

Double carboxylation reactions, in which two $CO_2$ are introduced simultaneously onto a C–C unsaturated compounds, are highly promising for the efficient synthesis of dicarboxylic acid derivatives. Regarding stoichiometric reactions, Mori and co-worker reported an elegant double carboxylation of 1,3-butadienes in the presence of a Ni(0) complex.[56] 1-(Trimethylsilyl)-3-alkyl-substituted allenes were used as the substrate in the first catalytic double carboxylation with a Ni catalyst.[47] Recently, we reported the Ni-catalyzed double carboxylation of internal alkynes with $CO_2$, affording highly versatile maleic anhydrides as the products (Scheme 6).[44] The reaction of 5-decyne (**4a**) was investigated using $Ni(acac)_2(bpy)$ (10 mol %, acac = acetylacetonate; bpy = bipyridine) as the catalyst, Zn powder (3.0 equiv) as the reducing reagent, and $MgBr_2$ (2.0 equiv) as the additive in the presence of molecular sieves 3 Å (MS 3 Å, powder) in *N,N*-dimethylformamide (DMF) at room temperature under 1 atm pressure of $CO_2$. Under these conditions, the double-carboxylated product **5a** was obtained in 74% GC yield. When Ni(acac)$_2$ was used as the catalyst (i.e., without the bpy ligand), the reaction failed. In the reaction, $MgBr_2$ and Zn powder were indispensable. Without MS 3 Å, the yield of product **5a** decreased slightly.

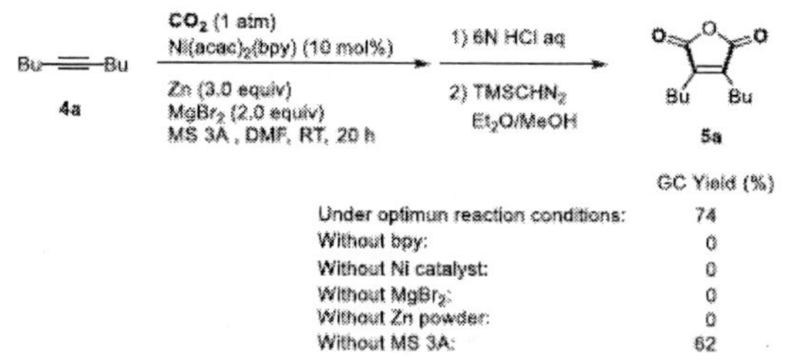


**Scheme 6.** Ni-catalyzed double carboxylation of **4a**

Under the optimal reaction conditions, **5a** was obtained in 71% isolated yield (Scheme 7). When 3-hexyne (**4b**) and 2,9-dimethyl-5-decyne (**4c**) were used as the substrates, products (**5b** and **5c**) were obtained in good isolated yields. 2-Octyne (**4d**) and 5-phenyl-2-pentyne (**4e**) reacted smoothly, affording the corresponding products. A cyclic internal alkyne (**4f**) also participated in the reaction, giving the product (**5f**) in 79% yield. An alkyne with a C=C bond (**4g**) was converted to the corresponding product (**5g**). Alkynes with secondary alkyl groups on the *sp*-carbons (**4h** and **4i**) also afforded the double-carboxylated products (**5h** and **5i**) in good yields. Unfortunately, terminal alkynes and aromatic internal alkynes did not afford the corresponding double-carboxylated products.

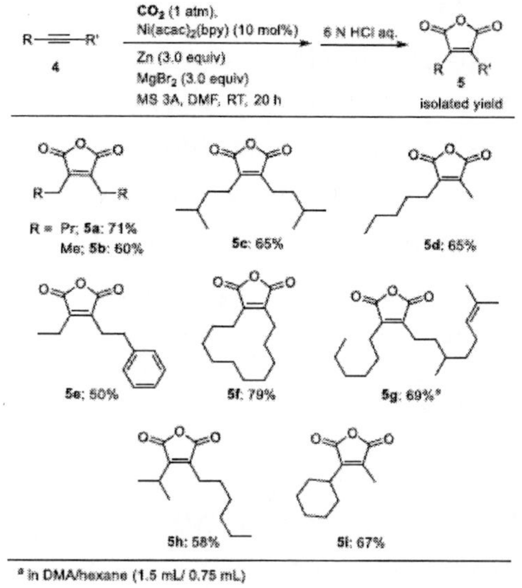


**Scheme 7.** Ni-catalyzed double carboxylation of alkynes (**4**)

Hydrocarboxylation of Alkynes

Copper hydride (CuH) is known as an efficient reducing agent in organic synthesis.[57-59] Various CuH catalyzed reductions of carbonyl compounds such as ketones and aldehydes have been reported employing hydrosilanes, which is an easy-to-handle and stable reducing agent. In contrast, the CuH species also adds across nonpolar C–C bonds, generating reactive organocopper intermediates.

Using the strategy, we found that Cu complexes catalyzed the hydrocarboxylation of alkynes using $CO_2$ and hydrosilane as the reducing agent.[39] The reaction of diphenylacetylene

(**6a**) with $CO_2$ (balloon) was carried out using $HSi(OEt)_3$ as the reducing agent in dioxane at 100 °C (Scheme 8). The yield of (*E*)-2,3-diphenyl-2-propenoic acid **7a**) was determined by GC after derivatization to the corresponding methyl ester (**7a**-Me). When IMesCuF was used as the catalyst, **7a**-Me was obtained in 86% yield. When IPrCuF, which is a bulkier catalyst compared to IMesCuF, was used as the catalyst, the yield of 7a-Me decreased to 41%. When $HSi(OEt)_3$ was replaced with polymethylhydrosiloxane (PMHS) that is an inexpensive, easy-to-handle, and environment-friendly reducing agent, **7a**-Me was obtained in 80% yield. Toluene was an acceptable solvent, affording **7a**-Me in 81% yield.

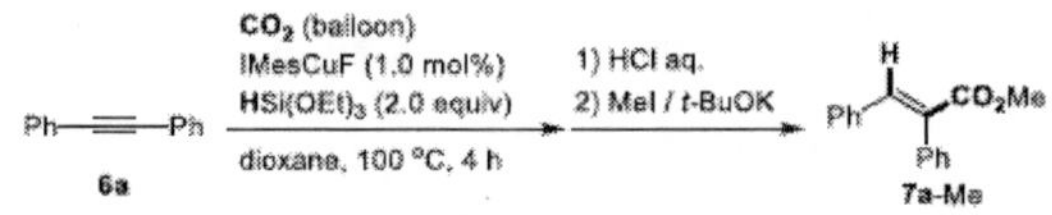


| | GC Yield (%) |
|---|---|
| Under optimun reaction conditions: | 86 |
| IPrCuF in place of IMesCuF | 41 |
| PMHS in place of $HSi(OEt)_3$ | 80 |
| $HSiEt_3$ in place of $HSi(OEt)_3$ | 0 |
| Toluene in place of dioxane | 81 |

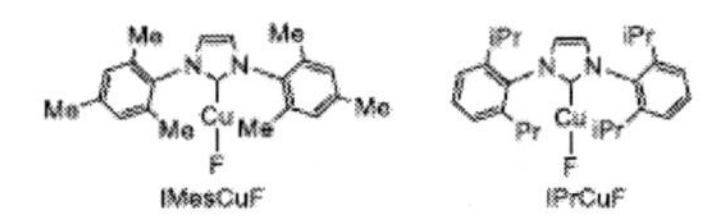


**Scheme 8.** Cu-catalyzed hydrocarboxylation of diphenylacetylene (**6a**)

The hydrocarboxylation of various symmetrical aromatic alkynes (**6b–6h**) was carried out in the presence of IMesCuF as the catalyst (Scheme 9). From all the alkynes listed, the corresponding $\alpha,\beta$-unsaturated carboxylic acids (**7b–h**) were obtained in good isolated yields with perfect (*E*) stereochemistry. The reaction of alkynes bearing both electron-rich and electron-deficient aryl moieties afforded the corresponding products in good yields. Importantly, bromo (**7d**), alkoxycarbonyl (**7e**) and cyano (**7f**) functional groups were tolerated in the reaction. As for terminal alkynes, phenylacetylene (**6i**) was converted to cinnamic acid (**7i**) in 44% yield when IPrCuF was used as the catalyst.

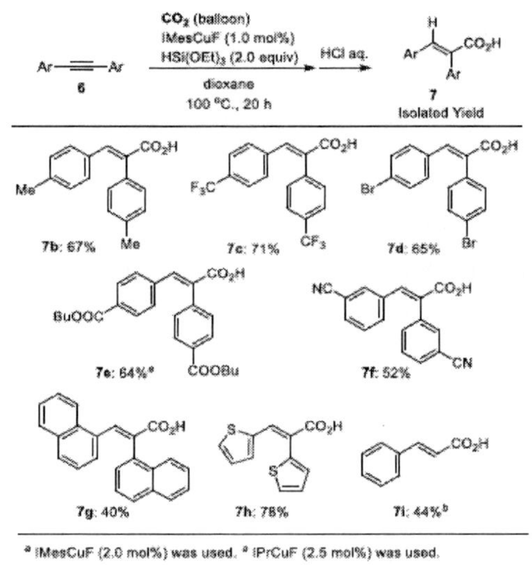


**Scheme 9.** Cu-catalyzed hydrocarboxylation of aromatic alkynes (**6**)

In the hydrocarboxylation of 1-phenyl-3,3-dimethyl-1-butyne (**6j**), when $Cl_2$IPrCuF (2.5 mol%) was used in hexane as the solvent at 70 °C, a single isomer (**7j**) was obtained in 71% yield (Scheme 10). The reaction of an alkyne with cyclohexyl groups (**6k**) afforded the corresponding product (**7k**) in high yield. In the case of 1-phenyl-1-hexyne as the substrate, a mixture of regioisomers (**7l**) was obtained in high yield and with good selectivity. Moreover, the reaction of 1,4-dimethoxy-2-butyne (**6m**) and 2,5-dimethoxy-3-hexyne (**6n**) bearing two propargylic ether functional groups also afforded the corresponding products (**7m** and **7n**) in good yields.

In order to gain insights into reaction mechanism, stoichiometric reactions were carried out (Scheme 11). When $Cl_2$IPrCuF was treated with an excess (4 equiv) of hydrosilanes such as PMHS or $HSi(OEt)_3$ in $C_6D_6$, an immediate color change from colorless to bright orange was observed. The $^1H$ NMR spectrum of the product indicated the formation of pure $Cl_2$IPrCuH with a proton resonance of Cu-H at 2.39 ppm, which was comparable to that for IPrCuH (2.67 ppm).[60] The reaction of an aromatic alkyne for 2.5 h at room temperature smoothly afforded the corresponding alkenyl-copper complex. The reaction with $CO_2$ (balloon) was very slow at room temperature, but proceeded at a higher reaction temperature (65 °C) for 12 h, giving a carboxylatocopper complex. Finally, the treatment with an excess (4 equiv) of $HSi(OEt)_3$ at room temperature smoothly afforded $Cl_2$IPrCuH.

Based on the stoichiometric reactions, a catalytic cycle is proposed as shown in Scheme 12. A CuH species (**A**) is generated *in situ* from LCuF (L = IMes, IPr, or $Cl_2$IPr) and a hydrosilane because of strong Si–F interactions (step a). The *syn* addition of **A** to an alkyne (**1**) initiates the catalytic cycle and affords a copper alkenyl intermediate (**B**) stereoselectively (step b). Then, the insertion of $CO_2$ gives the corresponding carboxylato copper intermediate (**C**) (step c). Finally, σ-bond metathesis of **C** with a hydrosilane provides the corresponding silyl ester and regenerates **A** (step d).

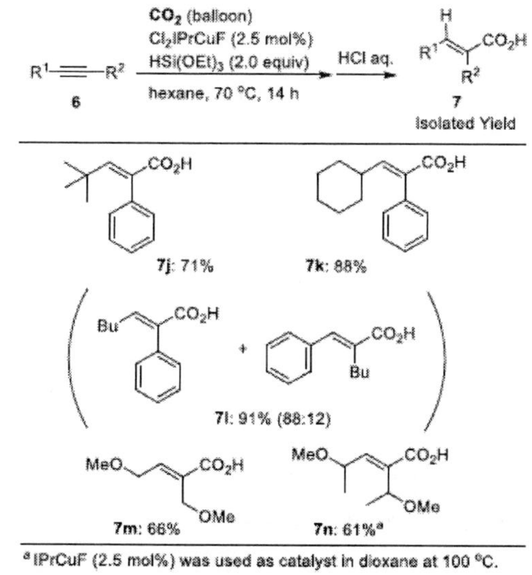


ᵃ IPrCuF (2.5 mol%) was used as catalyst in dioxane at 100 °C.

**Scheme 10.** Cu-catalyzed hydrocarboxylation of internal alkynes (**6**)

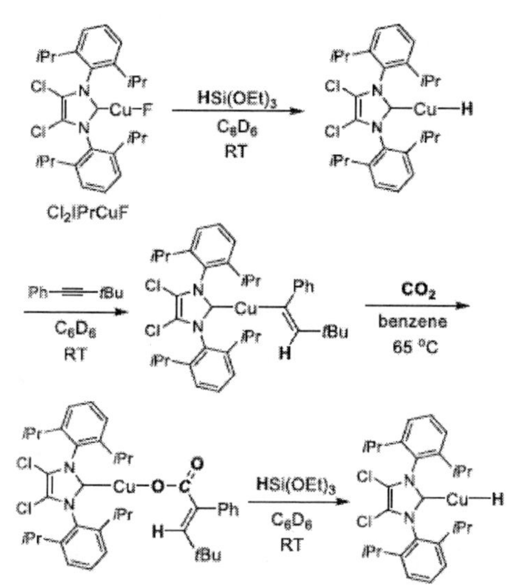


**Scheme 11.** Stoichiometric reactions relevant to mechanism employing $Cl_2$IPrCuF

## Silacarboxylation of Alkynes

Catalytic *hetero*carboxylation, in which the heteroatom and $CO_2$ are simultaneously and catalytically incorporated into unsaturated substrates, is extremely useful, because the reaction using $CO_2$ provides a valuable synthetic route for highly functionalized carboxylic acid derivatives. In this regard, Hou and co-workers reported the boracarboxylation of alkynes using Cu catalysts and bis(pinacolato)diboron as the boron source.[61]

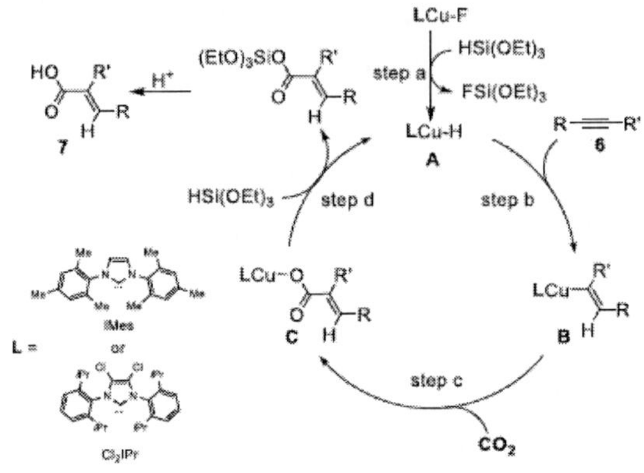

**Scheme 12.** Plausible reaction mechanism of Cu-catalyzed hydrocarboxylation of alkynes (**6**)

We reported the first catalytic silacarboxylation of internal alkynes employing $CO_2$ and silylborane in the presence of a Cu catalyst (Scheme 13).[62] The reaction of 1-phenyl-1-propyne (**8a**) was carried out using readily available $Me_2PhSi$-B(pin) as the silicon source in the presence of a Cu catalyst in octane at 100 °C under 1 atm pressure of $CO_2$. A mixture of CuCl and $PCy_3$ (P/Cu = 1) was used as the catalyst, affording silalactone (**9a**) in 97% yield and with high regioselectivity (**9a**/**9a′** = 97:3). P*t*$Bu_3$ also acted as an efficient ligand, and the products were obtained regioselectively in high yield. However, $PBu_3$ and $PPh_3$ were not efficient ligands. The reaction using IMesCuCl gave **9a** in high yield, but with slightly lower regioselectivity. $[CuCl(PCy_3)]_2$ prepared from CuCl and $PCy_3$ was the best catalyst, affording silalactone **9a** in a quantitative yield while maintaining high regioselectivity. When toluene and 1,4-dioxane were used as the solvent, the products were obtained in 98% and 84% yields, respectively, whereas the reaction using acetonitrile and DMF failed.

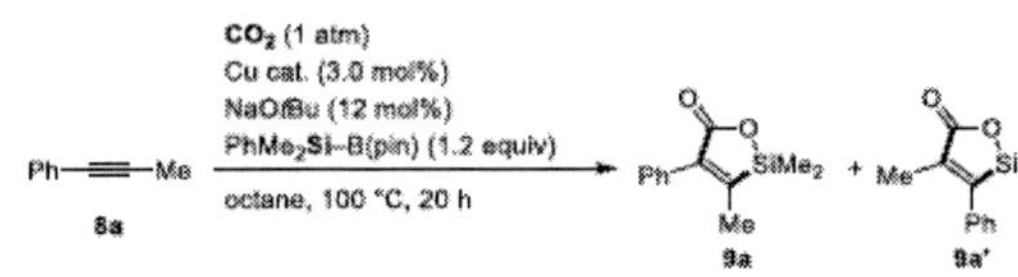

| Cu cat. | GC Yield | (9a/9a') |
|---|---|---|
| CuCl/PCy₃ | 97% | (97/3) |
| CuCl/P(tBu)₃ | 90% | (97/3) |
| CuCl/PPh₃ | 28% | (74/26) |
| IMesCuCl | 96% | (88/12) |
| IPrCuCl | 18% | (77/23) |
| [CuCl(PCy₃)]₂ (2.5 mol%) | 99% | (96/4) |

**Scheme 13.** Cu-catalyzed silacarboxylation of **8a**

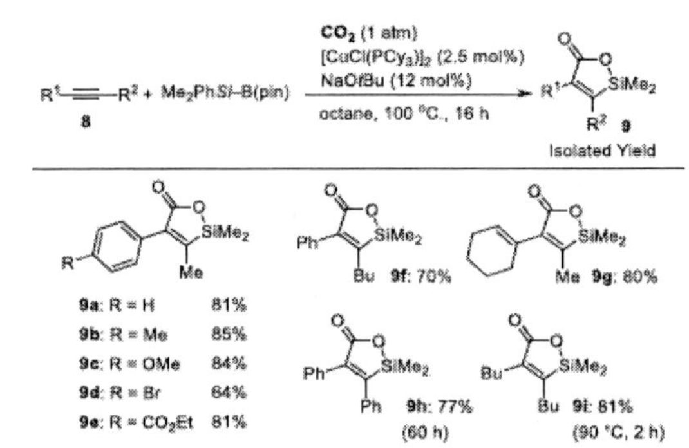

**Scheme 14.** Cu-catalyzed silacarboxylation of alkynes (**8**)

The reactions of 1-aryl-1-propynes bearing both electron-rich and electron-deficient substituents on the aryl ring afforded the corresponding silalactones **9b–e** (Scheme 14). Gratifyingly, the bromo (**9d**) and ester (**9e**) functional groups remained intact under these reaction conditions. The reactions of 1-phenyl-1-hexyne (**8f**) and a conjugated enyne (**8g**) afforded the corresponding products (**9f** and **9g**) in 70% and 76% yields, respectively. Diphenylacetylene (**8h**) also afforded the corresponding product (**9h**) in good isolated yield with a longer reaction time, whereas the reaction of 5-decyne (**8i**) at 90 °C for 2 h afforded **9i** in 81% yield.

Control experiments were carried out (Scheme 15) to elucidate the silacarboxylation mechanism. The reaction may proceed stepwise via two sequential known reactions, i.e., the silaboration of the alkyne followed by the carboxylation of the resulting vinylboronic ester.[32,33] We first prepared a possible intermediate **10i** by the reaction of **9i** with $Me_2PhSi$-B(pin) using a Pt catalyst. Then, the reaction of intermediate **10i** with $CO_2$ was carried out under the standard reaction conditions. **10i** did not react; thus silalactone **9i** was not obtained. Moreover, **10i** was not detected in the reaction mixture at a low conversion of **8i** (reaction time: 15 min; conversion of **8i**: 19%; yield of **9i**: 12%). Thus, these results clearly indicate that the silacarboxylation proceeds in one step, not via the silaboration of the alkyne followed by carboxylation.

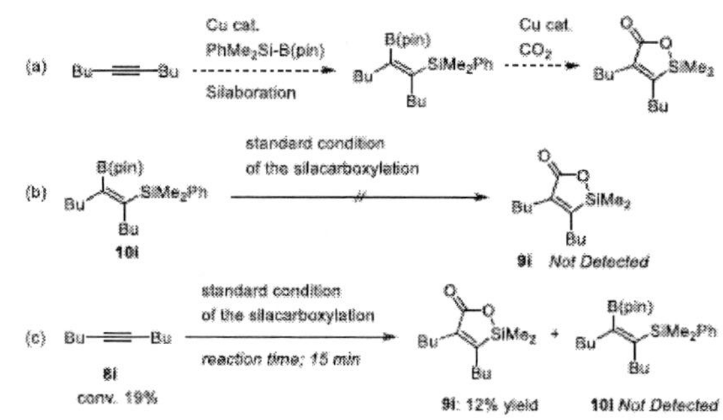


**Scheme 15.** Control experiments relevant to mechanism.

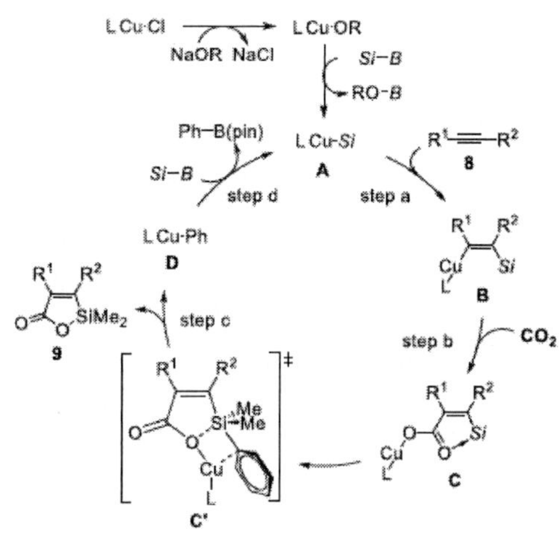


**Scheme 16.** Plausible reaction mechanism of copper-catalyzed silacarboxylation of alkynes (**8**)

Based on these observations, a possible catalytic cycle for the silacarboxylation of alkynes with $CO_2$ is proposed (Scheme 16). First, the reaction of the alkoxocopper species with silylborane affords the silylcopper complex (**A**). The *syn* addition of **A** to an alkyne **8** initiates the catalytic cycle and affords the $\beta$-silylalkenylcopper intermediate **B** (step a). Then, the insertion of $CO_2$ into the Cu–C(vinyl) bond gives the corresponding Cu carboxylato species (**C**) (step b). The intramolecular cyclization via **C´** affords the phenylcopper species **D**, providing the silalactone (**9**) (step c). Finally, the $\sigma$-bond metathesis of **D** with silylborane gives PhB(pin), and the silylcopper species (**A**) is regenerated (step d).

Recently, silacarboxylation of allenes[63] and carboxyzincation of alkynes[64] with $CO_2$ were also reported.

## References

[1] New and Future Developments in Catalysis, Activation of Carbon Dioxide, ed. by Suib, S. L., Elsevier, Amsterdam (2013).

[2] Carbon Dioxides as Chemical Feedstock, ed. by Aresta, M., Wiley-VHC, Weinheim (2010).

[3] Federsel, C., Jackstell, R., Beller, M., *Angew. Chem. Int. Ed.,* **49**, 6254 (2010). .

[4] Jessop, P. G., Joó, F., Tai C.-C., *Coord. Chem. Rev.*, **248**, 2425 (2004).

[5] Leitner, W., *Angew. Chem. Int. Ed.,* **34**, 2207 (1995).

[6] Jessop, P. G., Ikariya, T., Noyori, R. *Chem. Rev.*, **95**, 259 (1995).

[7] Martin, C., Fiorani, G., Kleij, A. W., *ACS Catal.*, **5**, 1353 (2015).

[8] Kember, M. R., Buchard, A., Williams, C. K., *Chem. Commun,*. 47, **141** (2011).

[9] Clegg, W., Harrington, R. W., North, M., Pasquale, R. *Chem. Eur. J.*, **16**, 6828 (2010).

[10] Decortes, A., Castilla, A. M., Kleij, A. W., *Angew. Chem. Int. Ed.*, **49**, 9822 (2010).

[11] Sakakura, T., Kohno, K., *Chem. Commun.*, 1312 (2009).

[12] Maeda, C., Miyazaki, Y., Ema, T., *Catal. Sci. Technol.*, **4**, 1482 (2014).

[13] Yeung, C. S., Dong, V. M., *Top. Catal.*, **57**, 1342 (2014).

[14] Cai, X., Xie, B., *Synthesis*, **45**, 3305 (2013).

[15] Zhang, L., Hou, Z. *Chem. Sci.*, 2013, **4**, 3395 (2013).

[16] Tsuji, Y., Fujihara, T., *Chem. Commun.*, **48**, 9956 (2012).

[17] Manjolinho, F., Arndt, M., Gooßen, K., Gooßen, L. J., *ACS Catal.*, **2**, 2014 (2012)

[18] Huang, K., Sun, C.-L., Shi, Z.-J., *Chem. Soc. Rev.*, **40**, 2435 (2011).

[19] Cokoja, M., Bruckmeier, C., Rieger, B., Herrmann, W. A., Kuhn, F. E., *Angew. Chem., Int. Ed.*, **50**, 8510 (2011).

[20] Riduan, S. N., Zhang, Y., *Dalton Trans.*, **39**, 3347 (2010).

[21] Wang, X., Liu, Y., Martin, R., *J. Am. Chem. Soc.*, **137**, 6476 (2015).

[22] Moragas, T., Cornella, J., Martin, R., *J. Am. Chem. Soc.*, **136**, 17702 (2014).

[23] Liu, Y., Cornella, J., Martin, R., *J. Am. Chem. Soc.*, **136**, 11212 (2014)

[24] Correa, A., León, T., Martin, R., *J. Am. Chem. Soc.*, **136**, 1062 (2014).

[25] León, T., Correa, A., Martin, R., *J. Am. Chem. Soc.*, **135**, 1221 (2013).

[26] Tran-Vu, H., Daugulis, O., *ACS Catal.*, **3**, 2417 (2013).

[27] Fujihara, T., Nogi, K., Xu, T., Terao, J., Tsuji, Y., *J. Am. Chem. Soc.,* **134**, 9106 (2012).

[28] Correa, A., Martin, R., *J. Am. Chem. Soc.,* **131**, 15974 (2009).

[29] Yeung, C. S., Dong, V. M., *J. Am. Chem. Soc.*, **130**, 7826 (2008).

[30] Ochiai, H., Jang, M., Hirano, K., Yorimitsu, H., Oshima, K., *Org. Lett.*, **10**, 2681 (2008).

[31] Kobayashi, K., Kondo, Y., *Org. Lett.*, **11**, 2035 (2009).

[32] Onishi, T., Nishiura, M., Hou, Z., *Angew. Chem., Int. Ed.*, **47**, 5792 (2008).

[33] Takaya, J., Tadami, S., Ukai, K., Iwasawa, N., *Org. Lett.*, **10**, 2697 (2008).

[34] Ukai, K., Aoki, M., Takaya, J., Iwasawa, N., *J. Am. Chem. Soc.*, **128**, 8706 (2006).

[35] Duong, H. A., Huleatt, P. B., Tan, Q.-W., Shuying, E. L., *Org. Lett.*, **15**, 4034 (2013).

[36] Ohishi, T., Zhang, L., Nishiura, M., Hou, Z., *Angew. Chem. Int. Ed.*, **50**, 8114 (2011).

[37] Ohmiya, H., Tanabe, M., Sawamura, M., *Org. Lett.*, **13**, 1086 (2011).

[38] Li, S., Yuan, W., Ma, S., *Angew. Chem., Int. Ed.*, **50**, 2578 (2011).

[39] Fujihara, T., Xu, T., Semba, K., Terao, J., Tsuji, Y. *Angew. Chem., Int. Ed.*, **50**, 523 (2011).

[40] Williams, C. M., Johnson, J. B., Rovis, T., *J. Am. Chem. Soc.*, **130**, 14936 (2008).

[41] Tani, Y., Kuga, K., Fujihara, T., Terao, J., Tsuji, Y., *Chem. Commun.*, **51**, 13020 (2015).

[42] Takaya, J.; Iwasawa, N. *J. Am. Chem. Soc.*, **130**, 15254 (2008).

[43] Takaya, J., Sasano, K., Iwasawa, N., *Org. Lett.*, **13**, 1698 (2011).

[44] Fujihara, T., Horimoto, Y., Mizoe, T., Sayyed, F. B., Tani, Y., Terao, J., Sakaki, S., Tsuji, Y., *Org. Lett.*, **16**, 4960 (2014).

[45] Ishii, M., Mori, F., Tanaka, K., *Chem. Eur. J.*, **20**, 2169 (2014)

[46] Shimizu, K., Takimoto, M., Sato,Y., Mori, M., *Org. Lett.*, **7**, 195 (2005).

[47] Takimoto, M., Kawamura, M., Mori, M., Sato, Y., *Synlett*, 2019 (2005).

[48] Louie, J., Gibby, J. E., Farnworth, M. V., Tekavec, T. N., *J. Am. Chem. Soc.*, **124**, 15188 (2002).

[49] Takimoto, M., Shimizu, K., Mori, M., *Org. Lett.*, **3**, 3345 (2001).

[50] Tsuda, T., Morikawa, S., Sumiya, R., Saegusa, T., *J. Org. Chem.*, **53**, 3140 (1988).

[51] Kishimoto, Y., Mitani, I., *Synlett*, 2141 (2005).

[52] Hidai, M., Kashiwagi, T., Ikeuchi, T., Uchida, Y., *J. Organomet. Chem.*, **30**, 279 (1971).

[53] Amatore, C., Jutand, A., *J. Am. Chem. Soc.*, **113**, 2819 (1991).

[54] Sayyed, F. B., Tsuji, Y. Sakaki, S., *Chem. Commun.*, **49**, 10715 (2013).

[55] Nogi, K., Fujihara, T., Terao, J., Tsuji, Y., *Chem. Commun.*, **50**, 13052 (2014).

[56] Takimoto, M., Mori, M., *J. Am. Chem. Soc.*, **123**, 2895 (2001).

[57] Díez-González, S., Nolan, S. P., *Acc. Chem. Res.*, **41**, 349 (2008).

[58] Rendler, S., Oestreich, M., *Angew. Chem. Int. Ed.*, **46**, 498 (2007).

[59] Deutsch, C., Krause, N., Lipshutz, B. H., *Chem. Rev.*, **108**, 2916 (2008).

[60] Mankad, N. P., Laitar, D. S., Sadighi, J. P., *Organometallics*, **23**, 3369 (2004).

[61] Zhang, L., Cheng, J., Carry, B., Hou, Z. *J. Am. Chem. Soc.*, **134**, 14314 (2012).

[62] Fujihara, T., Tani, Y., Semba, K., Terao, J., Tsuji, Y., *Angew. Chem. Int. Ed.*, **51**, 11487 (2012).

[63] Tani, Y., Fujihara, T., Terao, J., Tsuji, Y., *J. Am. Chem. Soc.*, **136**, 17706 (2014).

[64] Nogi, K., Fujihara, T., Terao, J., Tsuji, Y., *J. Am. Chem. Soc.*, **138**, 5547 (2016).

# Paper No. 16

# Catalyst Development for hydrogenation of $CO_2$ for energy storage

***Ali Aljishi[1*], Gabriel Veilleux[2], Jose Augusto Hernandez Lalinde[2], Jan Kopyscinski[2], Emad N Al-Shafei[1]***
[1] *Researach and Development Center, Saudi Aramco, Dhahran,*
[2] *Mcgill Univerisisty, Montreal, Canada*

Renewable energy sources have been developed to provide more sustainable and additional environmental way to generate power and to overcome the $CO_2$ emissions due to the burning of fossil fuels. However, the utilization of the renewables are challenging since they cannot produce electricity on demand compared to the conventional power plants which creates a lot of variations on the electric grids and on the load levels. To overcome the variation problems, Power-to-Gas technology has been proposed, in which electrical energy is converted into chemical energy and stored in the natural gas grid. This technology has higher storage capacity and discharge time than technologies currently in use such as batteries and pumped hydro storage.

In Power-to-Gas technology, methanation is the most important and challenging step, in which $CO_2$ is reacted with excess of $H_2$ over a catalyst to produce methane. The main overall reactions are the $CO_2$ methanation (eq. 1), the competing reverse water-gas-shift reaction (eq. 2), which leads to undesired CO, and the subsequent CO methanation (eq.3).

$$CO_2 + 4H_2 \rightarrow CH_4 + 2H_2O \qquad \Delta H_R^o = -165\ kJ\ mol^{-1} \qquad (1)$$

$$CO_2 + H_2 \leftrightarrow CO + H_2O \qquad \Delta H_R^o = 41\ kJ\ mol^{-1} \qquad (2)$$

$$CO + 3H_2 \leftrightarrow CH_4 + H_2O \qquad \Delta H_R^o = -206\ kJ\ mol^{-1} \qquad (3)$$

Methanation reactions are extremely exothermic as this affect and deactivate the catalyst. The current work focusses on developing nickel containing ordered mesoporous alumina (OMA) catalysts for $CO_2$ methanation. Ordered mesoporous materials are one of the most promising catalysts nowadays due to their unique properties such as high surface area, large pore volume, ordered and amorphous structure. The strong interaction between the metal oxide and the support lead to high metal dispersion, which results in better catalytic activity and stability at high reaction temperatures. Evaporation Induced Self Assembly (EISA) technique was used to synthesize a series of ordered mesoporous nickel alumina catalysts. Catalyst morphology and activity was significantly influenced by changing synthesis parameters such as acid type, nickel loading, calcination temperature as well as synthesis method.

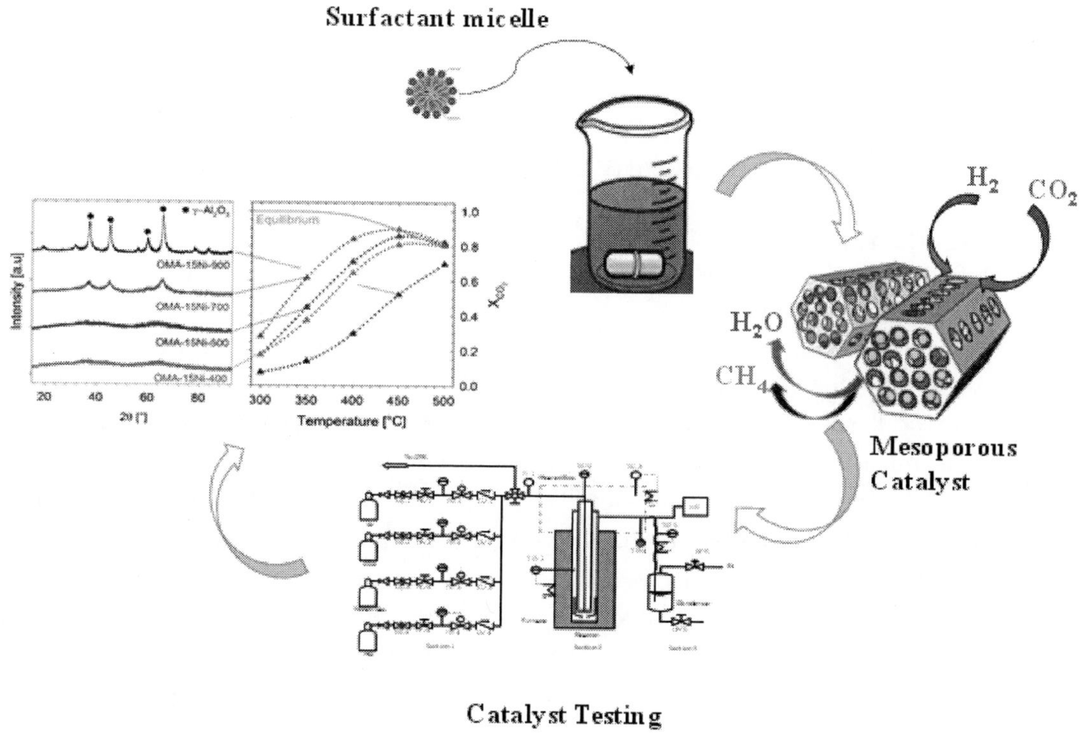


The absence of acidic media in the synthesis produced a catalyst with macroporous structures and low surface area (47 $m^2\ g^{-1}$), whereas using a mixture of hydrochloric acid and citric acid led to an incomplete formation of mesoporous micelles with a surface area of 173 $m^2\ g^{-1}$. A complete formation of long cylindrical micelles with a combined surface areas up to 260 $m^2\ g^{-1}$ and highly dispersed nickel clusters with a size of 3-5 nm was achieved via the usage of nitric acid.

This work investigated the influence of the catalyst synthesis on the catalyst properties and reactivity of hydrogenation reaction. Catalysts were characterized using BET, TPR, XRD, TEM, and ICP, furthermore, catalytic activity towards $CO_2$ hydrogenation was tested through fixed bed reactor at different conditions in order to meet equilibrium hydrogenation of $CO_2$. An optimum calcination temperature of 700 °C was determined yielding the highest $CO_2$ conversion and $CH_4$ selectivity. This catalyst showed a stable performance and did not display any sign of deactivation during 150 h time-of-stream.

Paper No. 17

# Size-Controlled Synthesis of Gold Nanoparticles on Fibrous Silica Nanospheres (KCC-1) and their activity in the oxidation of CO

***Ziyauddin S. Qureshi***
*Center of Research Excellence in Petroleum Refining and Petrochemicals, KFUPM, Dhahra,*
*Email: zqureshi@kfupm.edu.sa*

Size and shape controlled synthesis of highly dispersed nanoparticles remains a major challenge in the research on nanoparticles even after the development of several methods for their preparation. By tuning the size and shape of a nanoparticle, the intrinsic properties of the nanoparticle can be controlled leading tremendous potential applications in different fields of science and technology. We describe a facile route for the synthesis of gold nanoparticles (Au NPs) of different average size (ranging from 1-2 nm for Au/KCC-1-$NH_2$-a1 to 11-13 nm for Au/KCC-1-$NH_2$-b1) on to the fibrous silica nanospheres (KCC-1) either by immobilizing pre-made Au NPs on Au/KCC-1-$NH_2$ or by grafting $HAuCl_4$ on Au/KCC-1-$NH_2$ followed by reduction. The size and the location of the Au NPs on the support were found to depend on the preparation method. The KCC-1-supported Au NPs were thoroughly characterized by using HR-TEM, high-angle annular dark field scanning transmission electron microscopy (HAADF-STEM), XRD, X-ray photoelectron spectroscopy, UV, and Brunauer-Emmett-Teller surface area measurements and were applied in catalysis for the oxidation of CO. The catalytic activity of the Au NPs over KCC-1 is proved to be size dependent.

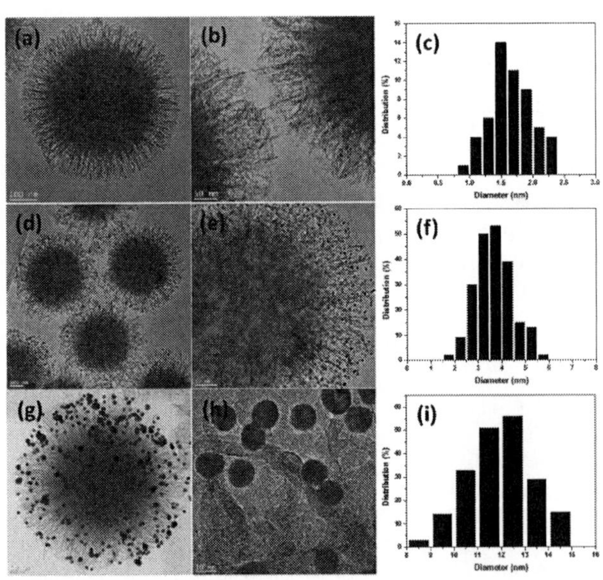


**Figure 1.** HR-TEM images and size histogram of Au/KCC-1-$NH_2$-a1 (a–c), Au/KCC-1-$NH_2$-a2 (d–f), and Au/KCC-1-$NH_2$-b1 (g–i).

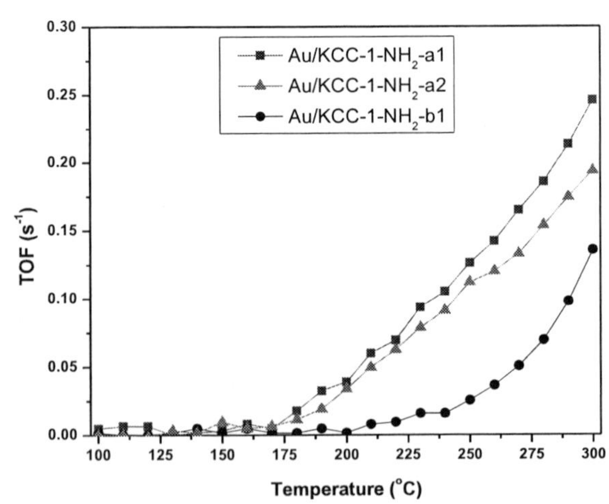


**Figure 2.** Turn over frequency (TOF $S^{-1}$) for CO oxidation over Au/KCC-1-$NH_2$-a1, Au/KCC-1-$NH_2$-a2 and Au/KCC-1-$NH_2$-b1 after $H_2$ treatment at 300 °C for one hour.

Paper No. 18

# Bifunctional Selective Hydrosilylation of Terminal Alkynes by Bis(carbene) Iridium(III) and Rhodium(III) Complexes.

***Luis A. Oro***
*Departamento de Química Inorgánica, Instituto de Síntesis Química y Catálisis Homogénea, Universidad de Zaragoza-CSIC, 50010-Zaragoza, España / CoRE Petroleum Refining & Petrochemicals, KFUPM Visiting Chair Professor, Dhahran, 31261 E-Mail: oro@unizar.es*

Transition metal-catalyzed hydrosilylation of terminal alkynes is an efficient and atom economic route to vinylsilanes, and the selective formation of (Z)-vinylsilanes, or vinylsilanes is of relevant interest.

In the search for selective catalytic hydrosilylation of terminal alkynes, we have prepared a family of chelating bis(carbene) cationic complexes of formulae $[M(I)_2\{\kappa$-C,C,O,O-bis(NHC)$\}]BF_4$ (M= Ir, Rh; bis-NHC = methylenebis(N-2-methoxy- ethyl)imidazole-2-ylidene) or $[Rh(XY)_2\{\kappa$-C,C,O,O-bis(NHC)$\}]BF_4$ (X = $CF_3CO_2$, Y = I or $CF_3CO_2$). Interestingly, $[M(I)_2\{\kappa$-C,C,O,O-bis(NHC)$\}]BF_4$ complexes showed to be very effective for the hydrosilylation of a range of terminal alkynes, employing different silanes. The reactions proceed with excellent yields and selectivities to their corresponding ⍰-(Z)-vinylsilanes, but only when acetone was used as solvent. In general, the iridium catalyst is significantly more active than the rhodium one for the hydrosilylation of aromatic alkynes, whereas the trend observed for aliphatic alkynes is the opposite.

We propose that the specificity of acetone for the success of the hydrosilylation reaction, catalyzed by $[M(I)_2\{\kappa$-C,C,O,O-bis(NHC)$\}]BF_4$ complexes, is based on Si-O interactions between the silane and the acetone solvent, favouring a metal-ligand bifunctional outer-sphere mechanism. The proposed stepwise outer-sphere mechanism is based on the initial heterolytic splitting of the silane molecule by the metal centre in such way that the acetone molecule facilitate the transfer of the $R_3Si^+$ moiety from the silane to the terminal alkyne by means of an oxocarbenium ion, as previously proposed by Park and Brookhart on ketone hydrosilylation by an electrophilic $\eta^1$-silane iridium(III) complex. In the next step the resulting oxocarbenium ion ($[R_3Si{-}O(CH_3)_2]^+$) reacts with the corresponding alkyne to give the silylation product ($[R_3Si{-}CH{=}C{-}R]^+$), followed by the selective nucleophilic attack of the hydrido ligand over $[R_3Si{-}CH{=}C{-}R]^+$. The excellent ⍰-(*Z*)-selectivity of the reaction could be explained as a consequence of the higher steric interaction resulting from the geometry of the approach that leads to ⍰-(*E*)-vinylsilanes. The mechanism proposal represents the first example of an outer-sphere mechanism for the hydrosilylation of terminal alkynes, in which the acetone acts as a silane shuttle transferring the silyl moiety from the silane to the alkyne.

Paper No. 19

# Process Intensification, a promising approach in separating a ternary system using distillation to reduce energy consumption

***Dafer Al-Shahrani, Jagan Rallapalli***
*SABIC R&D, Riyadh*

It is obvious that distillation is employed for more than 90% of process separations in the process industries. It has been estimated that about 3% of the world energy consumption is utilized for distillation separation only. Though distillation is a pivotal process in process industries, it is also an energy intensive. Hence, the need for new optimized distillation sequences, tight and thermal integration of distillation columns, and optimized operation of distillation in terms reflux rate, boil-up ratio is paramount in the present world scenario.

Motivated by the large energy requirements for distillation, process intensification approach, which is a promising design philosophy that integrates distinct processes or minimizes no. of operations and equipment, is chosen to reduce the energy consumption in distillation processes. The process Intensification technologies can also reduce capital investments, and achieve environmental and safety benefits as well.

In this study, benzene, toluene, and xylene ternary system is chosen, and a systematic and simultaneous optimization of the distillation separation system is performed to get minimum energy consumption with less capital cost. The optimization is carried out in terms of distillation sequence, process intensification, pressure, temperature & composition of the column, no.of stages, and feed tray location. The extent of the energy reduction depends on the composition of the feed as well. The study developed an optimized distillation separation method using Process Intensification (PI) approach, which consumes 15% less energy than that of presently employed conventional distillation method for the benzene, toluene, and xylene ternary mixture of equimolar molar composition. This new method also requires less capital cost for a green field project. The study also advises general heuristics to employ a particular PI distillation sequence for separating a ternary mixture system. The methodology employed in the study, can also be utilized to optimize four-component mixture systems.

Paper No. 20

# Thermo-Neutral Reforming of Diesel for Auxiliary Power Unit (APU) Application

***Aadesh Harale[1], Mohammed Albuali[1], Ahmed Naimi[1], Mohamed Draze[1], Sai P. Katikaneni[1], Shakeel Ahmed[2]***
*[1]Research & Development Center, Saudi Arabian Oil Company (Saudi Aramco), Dhahran*
*[2]Center for Refining and Petrochemicals, Research Institute, King Fahd University of Petroleum & Minerals, Dhahran*

There is a significant interest in fuel cell technology applications due to environmental legislations for CO2 and other greenhouse gases emissions that demand the use of high efficiency energy production systems such as solid oxide fuel cell (SOFC) technology. Fuel cells are characterised by high operation efficiency, which results in decreased fuel consumption, and low environmental impact. A fuel cell is a device that converts the chemical energy of a fuel directly into electricity through electrochemical reactions, with significant waste heat making the overall system very efficient. SOFC technology has great potential of its application as auxiliary power unit (APU) for heavy duty vehicles during idling applications. Diesel based SOFC APU units are expected to be about 15-20% more efficient compared to commercially available diesel based APUs resulting in significantly lower CO2 footprint and lower emissions of criteria pollutants with almost no noise.

Thermo-neutral reforming (TNR) of diesel to produce hydrogen-rich syngas can be a viable source of fuel for solid oxide fuel cell based auxiliary power unit (APU). The fuel reforming process, which is integrated with a SOFC for APU application, plays an important role in APU system. The reformate gas consisting mainly of hydrogen, carbon monoxide, water and carbon dioxide is generated with diesel fuel reforming. The reformate produced will be sent to SOFC stack for electrochemical oxidation operating at temperatures around 800°C. However, diesel reforming has major challenges due to its high carbon content, aromatics, and sulphur, which play a major role in catalyst deactivation. The presentation will focus on ongoing diesel reforming catalyst development research work at Saudi Aramco Research and Development Center in collaboration with KKFUPM.

*27th Annual Saudi-Japan Symposium – 2017*

ارامكو السعودية
Saudi Aramco

# Technology in Petroleum Refining & Petrochemicals

*Venue: KFUPM Research Institute Bldg. 15; 4th Floor Auditorium, Dhahran, Saudi Arabia, November 20-21, 2017*

## Day One: Monday, November 20, 2017

**OPENING REMARKS SESSION** — ***Chairman: Dr. Sulaiman Al-Khattaf***

***8:15 Registration & Coffee Break***
***8:30 Opening Remarks***

- ***H.E. Dr. Khaled S. Al-Sultan**, Rector, KFUPM*
- ***Mr. Eiji Hiraoka**, Senior Executive Director, JCCP*
- ***Mr. Jun Mutoh**, President, JPI*

**SESSION ONE ADVANCES IN REFINING/FCC** — ***Chairman: Dr. Bandar Al-Solami***

| | | |
|---|---|---|
| **9:00** | **1.** | **Vision of the Japanese oil industry & importance of the education of human resources** *Mr. Jun Mutoh, JPI (Keynote)* |
| **9:30** | **2.** | **Commercialization of High-Severity fluid catalytic cracking (HS-FCC) process for light olefins production in refineries,** *Mr. Iwao Ogasawara, JXTG Energy, Japan* |
| **9:50** | **3.** | **A case study of HS-FCC reactor using kinetic modeling,** *Dr. Qi Xu, Saudi Aramco R&DC* |
| **10:10** | **4.** | **Degradation of heavy-duty coating applied to inner bottom plates of oil storage tanks and its evaluation method based on electrical impedance analysis,** *Dr. Shinji Okazaki, Yokohama National University, Japan* |
| **10:30** | | ***Coffee Break*** |

**SESSION TWO HYDROPROCESSING** — ***Chairman: Dr. Keiichi Tomishige***

| | | |
|---|---|---|
| **11:00** | **5.** | **Applications of process and kinetic models to optimize the operation of hydrocracker** *Dr. Rashid Ansari, Saudi Aramco R&DC* |
| **11:20** | **6.** | **Effect of acid medium and initial Si/Al ratio on the synthesis of mesoporous materials with enhanced acidity and hydrothermal stability,** *Dr. Lianhui Ding, Saudi Aramco R&DC* |
| **11:40** | **7.** | **Kinetics of simultaneous HDS of DBT and 4-MDBT/4,6-DMDBT over CoMoP/γ-$Al_2O_3$ catalysts,** *Dr. Syed A. Ali, KFUPM/RI* |
| ***12:00*** | | ***Prayer & Lunch Break*** |

**SESSION THREE UPGRADING AND REFORMING** — ***Chairman: Dr. Elie Fayad***

| | | |
|---|---|---|
| **13:00** | **8.** | **Development of a supercritical water cracking process for pipeline transportation of extra heavy crude oil,** *Mr. Takayoshi Fujimoto, JGC Corporation, Japan* |
| **13:20** | **9.** | **Hydrodynamics and solids mixing in fluidized beds with inclined-hole distributors** *Mr. Alhussain Bakhurji, Saudi Aramco R&DC* |
| **13:40** | **10.** | **Zeolite-based catalyst performance for catalytic reforming** *Mr. Mohammad Al-Rebh, Saudi Aramco R&DC* |
| ***14:00*** | | ***Day One Ends*** |

***Each presentation includes 5-minutes Q&A*** — **http://catsymp.kfupm.edu.sa/**

*27th Annual Saudi-Japan Symposium – 2017*

أرامكو السعودية
Saudi Aramco

# Technology in Petroleum Refining & Petrochemicals

*Venue: KFUPM Research Institute Bldg. 15; 4th floor Auditorium, Dhahran, Saudi Arabia, November 20-21, 2017*

## Day Two: Tuesday, November 21, 2017

**SESSION FOUR PETROCHEMICALS** — ***Chairman: Dr. Yasushi Tsuji***

*8:45* ***Registration & Coffee Break***

**9:00 11. Petrochemicals - technical and economic challenges for this century**
*Mr. Patrick Whitchurch, Honeywell UOP, USA (Keynote)*

**9:30 12. Catalytic copolymerization of $CO_2$ and diols using $CeO_2$ and 2-cyanopyridine**
*Dr. Keiichi Tomishige, Tohoku University, Japan*

**9:50 13. Toluene methylation to produce para-Xylene using ZSM-5 catalysts**
*Dr. Arudra Palani, KFUPM-RI; Dr. Ali Jahel, Honeywell UOP*

**10:10 14. Metathesis of 2-pentene over Mo and W supported mesoporous molecular sieves**
*Dr. M. Naseem Akhtar, KFUPM-RI*

**10:30** *Coffee Break*

**SESSION FIVE $CO_2$ UTILIZATION** — ***Chairman: Dr. Mohammed Ba-Shammakh***

**11:00 15. $CO_2$ fixations with homogeneous catalysis,** *Dr. Yasushi Tsuji, Kyoto University, Japan*

**11:20 16. Catalyst development for hydrogenation of $CO_2$ for energy storage**
*Mr. Ali Al-Jishi, Saudi Aramco R&DC*

**11:40 17. Size-controlled synthesis of gold nanoparticles on fibrous silica nanospheres (KCC-1) and their activity in the oxidation of CO,** *Dr. Ziyauddin Qureshi, KFUPM-RI*

*12:00* ***Prayer & Lunch Break***

**SESSION SIX HYDROSILYLATION/PROCESS STUDIES** — ***Chairman: Dr. Ali Al-Shareef***

**13:00 18. Bifunctional selective hydrosilylation of terminal alkynes by bis(carbene) iridium(III) and rhodium(III) complexes,** *Dr. Luis Oro, KFUPM/Zaragoza University, Spain*

**13:20 19. Process Intensification, a promising approach in separating a ternary system using distillation to reduce energy consumption,** *Mr. Dafer Al-Shahrani, SABIC R&D, Riyadh*

**13:40 20. Thermo-neutral reforming of diesel for auxiliary power unit (APU) application**
*Mr. Mohammed Albuali, Saudi Aramco R&DC*

***14:00 Closing Remarks, Symposium Ends***

*Each presentation includes 5-minutes Q&A* — ***Program as of Oct. 30, 2017***

*For Registration:* ***http://catsymp.kfupm.edu.sa/register.htm***